Estuaries: A Physical
Introduction

Estuaries: A Physical Introduction

K. R. Dyer

Unit of Coastal Sedimentation,
Taunton, Somerset

Visiting Lecturer, Department of Oceanography,
University of Southampton

A Wiley–Interscience Publication

JOHN WILEY & SONS

London · New York · Sydney · Toronto

Library of Congress catalog card number 72-8598

ISBN 0471 22905 9

Printed in Great Britain at The University Press, Aberdeen

Foreword

Estuaries occur in many varied forms, ranging from the coastal plain type to steep-sided fjords, but they all share the common feature of being regions where rivers and sea water meet and interact. Along many coasts the interaction is made more complex by the action of tides and tidal streams, the range of which is often increased when they penetrate an estuary. To the hydrologist and physical oceanographer the water movements and the turbulent mixing which results present interesting and challenging problems in hydrodynamics. These features, moreover, largely control the transport of material in suspension and its erosion or deposition: topics with which the geologist is concerned. Together with the chemical composition of the inflowing water, these characteristics lead to the distinctive features of the estuarine environment which make it of special interest to the biologist.

Many of the world's seaports are situated on estuaries and access to them depends on maintaining navigable channels of sufficient depth, a task which becomes increasingly difficult with the use of larger ships and the demand for improved docking facilities. The presence of a port is usually associated with centres of industry and population and for centuries it has been customary for their sewage and other wastes to be disposed of directly into the estuary. By reason of the relatively rapid mixing between estuary water and the open sea, enhanced by tidal action, many estuaries have coped more or less successfully with this imposition in the past but, with the increasing volume of waste material and the introduction of new and persistent chemical compounds, much more attention will have to be given in future to the conditions under which wastes are discharged into estuaries. At the same time the estuary is often required to act as a source of water supply for industry or for the cooling of power stations. In some areas schemes for the abstraction of water or the control of river discharge for irrigation purposes modify considerably the natural river flow and its incidence in time.

There are, therefore, a number of good reasons, both scientific and practical, for intensifying the study of estuaries and for disseminating more widely the knowledge which is acquired. This book is a contribution to that end, in that it describes and explains the physical processes acting in estuaries, to the extent that they are understood at present. It should be of interest to geologists and biologists and to those concerned with the management of estuary resources as well as to oceanographers and engineers. As well as dealing with the basic principles, the author discusses in detail their application to particular estuaries, drawing in several cases on his own varied experience in such investigations.

K. F. BOWDEN

January 1973

Preface

This book arose from a series of lecture notes compiled for a course on Estuaries given to a post-graduate class in the Oceanography Department, Southampton University. The course included students with first degrees in Zoology, Chemistry and Geology as well as Physics and Mathematics and, consequently, was a fairly low-level introduction into how estuaries work. As there was no complete textbook on the physics of estuaries the notes were polished and extended, but I hope they will still perform the function of introducing the relevant physical processes to all disciplines.

I think that one of the most interesting things about estuaries is that they are so complicated. They are continually in motion with cycles of variation that may never be repetitive and there is such close interlinking between the physical, chemical, geological and biological systems that research workers cannot afford to be too specialized in their consideration of estuaries. Each discipline must be aware of the needs of others. This is especially so regarding the physical oceanographers who can so easily work in isolation, but whose work is essential to others who depend on their quantifying the processes which govern the distribution of suspended and dissolved constituents. In return the physical oceanographers can achieve more if the other disciplines can formulate within realistic bounds what physical input they require. Perhaps this book will help disperse the state of the estuarine art to all those interested.

I would like to thank Professors K. F. Bowden and H. Charnock for their very helpful criticisms on the manuscript and Drs. J. Hinwood and P. A. Taylor for helpful discussions and advice. I also wish to thank Mrs. Bernardette Walling and Mrs. Joan Wedge for the typing and my wife and family for their continual encouragement.

'The search for truth is in one way hard and in another easy. For it is evident that no one can master it fully or miss it wholly. But each adds a little to our knowledge about Nature, and from all the facts assembled there arises a certain grandeur.' (Aristotle)

K. R. DYER

November 1972

Contents

Notation

A	Cross-sectional area of estuary
$\overline{A}$	Mean cross-sectional area
A_o	Tidal amplitude, Amplitude of tidal fluctuation in cross-sectional area
A	Tidal fluctuation of cross-sectional area
a	Pipe radius
B	A constant
b	Estuary breadth
C	A constant, Concentration of pollutant
C_o	Concentration of pollutant in segment at outfall
c	Wave celerity $= \sqrt{gh}$
D	A depth
D	Thickness of salt wedge
D	Dynamic depth
F	Flushing number
F	Density current transport
F	Flux of pollutant or salt through cross-section
F_i	Interfacial Froude number
F_m	Densimetric Froude number
f	Fresh water fraction, Frequency
f	Coriolis Parameter, $f_1 = 2\omega \sin \phi, f_2 = 2\omega \cos \phi$
G	Rate of energy dissipation per unit mass of water
g	Gravitational acceleration
H	A depth
h	Estuary depth
J	Rate of gain of potential energy
K_x	Coefficient of longitudinal eddy–diffusion
K_y	Coefficient of lateral eddy–diffusion
K_z	Coefficient of vertical eddy–diffusion
K_q	Coefficient of eddy–diffusion for pollutant
k	Wave number $k = 2\pi/\lambda = 2\pi n/u$, Coefficient of mortality, Coefficient of friction
k_o	Von Karman Constant $= 0\cdot4$
L	Length of salt wedge, Length of estuary, Amplitude of pollutant dispersal
l	Length scale of diffusion
M	Tidal mixing parameter

N_x Coefficient of longitudinal eddy-viscosity
N_y Coefficient of lateral eddy-viscosity
N_z Coefficient of vertical eddy-viscosity
n A number, Frequency
P Tidal prism volume, A factor
P Ratio of surface to r.m.s. tidal velocity
P Rate of introduction of a pollutant
p Pressure
Q Volume rate of transport of salt per unit width, Volume rate of transport of water
Q Amount of water in a section of an estuary
R River discharge
R_{10} River discharge averaged over ten days
Ra Estuarine Rayleigh number
Re Reynolds number
Ri Richardson number
Ri_E Estuarine Richardson number
r Exchange ratio
r Radius of curvature of streamlines
S Tidal fluctuation in salinity
S_o Mean salinity in segment, Amplitude of tidal fluctuation of salinity
S_n Mean salinity in segment n of estuary
S_s Salinity of undiluted sea water
s Salinity, Subscripts, superscripts etc., for s the same as for u
T Tidal period, Period of Turbulent oscillation, Tidal range
T Dimensionless wind stress
T Flushing time
t Time, Time constant
U Tidal variation of longitudinal velocity
U_o Amplitude of tidal variation of longitudinal velocity
U_* Friction velocity $= \sqrt{\tau_o/\rho}$
u Longitudinal velocity
u_o Observed longitudinal velocity averaged over a period of minutes
u' Turbulent velocity fluctuations of period less than a few minutes
$\bar{u}$ Tidal mean velocity
$\langle\bar{u}\rangle$ Tidal mean velocity averaged over depth
$\langle u\rangle$ Depth mean velocity
u_1 Deviation of longitudinal velocity from depth mean velocity
u_2 Deviation of depth mean velocity from cross-sectional mean
u_A Velocity averaged over cross-section
u_c Critical velocity for entrainment
u_d Deviation of velocity from cross-sectional average
u_f Fresh water velocity $= R/\bar{A}$

u_s Net surface current
u_t Root mean square tidal current
V   Tidal variation of lateral velocity
V   Low tide volume of segment of estuary
v   Lateral velocity, Subscripts, superscripts etc for v the same as for u
W   Tidal variation in vertical velocity
W_e Vertical velocity of entrainment
w   Vertical velocity, Subscripts, superscripts, etc. for w the same as for u
x Longitudinal distance
y Lateral distance
z Vertical distance
z_o Bed roughness length

α Specific volume $= 1/\rho$
δS Surface to bottom difference in tidal mean salinity
ε Molecular diffusion coefficient for salt
ζ Elevation of water surface
η Dimensionless vertical co-ordinate $= z/h$, Ratio between lateral and vertical shearing stresses
Θ Criterion of mixing
θ A phase lag, An angle
κ Constant in equation of state for sea water
λ Wavelength, A dimensionless length
μ Damping coefficient, Coefficient of molecular viscosity
ν Kinematic viscosity $= \mu/\rho$, Diffusive fraction of upstream salt flux
ξ Dimensionless horizontal co-ordinate
ξ_o Amplitude of horizontal tidal displacement
ρ Water density
ρ_f Density of fresh water
ρ_s Density of surface water
σ_{tH} Time of high water relative to high water at estuary head
σ_{rc}^2 Variance of dye distribution
τ Shearing stress
τ_o Shearing stress at the bed
ϕ Dissipation constant, Angle of latitude
ψ Stream function
ω Angular frequency of tide, Angular velocity of earth's rotation

CHAPTER 1

Introduction

An estuary can be defined in a variety of ways depending on one's immediate point of view. As far as most oceanographers, engineers and natural scientists are concerned, estuaries are areas of interaction between fresh and salt water. Consequently the definition most commonly adopted is that of Cameron and Pritchard (1963) who state that 'An estuary is a semi-enclosed coastal body of water which has a free connexion with the open sea and within which sea water is measurably diluted with fresh water derived from land drainage'. In this case we are restricting our interest to what have been termed positive estuaries (Pritchard, 1952b). A positive estuary is an estuary where the fresh water inflow derived from river discharge and precipitation exceeds the out-flow caused by evaporation. Surface salinities are consequently lower within the estuary than in the open sea. Negative estuaries are those where evapora-tion exceeds river flow plus precipitation and hypersaline conditions exist, e.g. Laguna Madre in Texas. Most estuaries are positive ones and we will confine our attention to those.

The interaction of fresh and salt water provides a circulation of water and mixing processes that are driven by the density differences between the two waters. The density of sea water depends on both the salinity and temperature, but in estuaries the salinity range is large and the temperature range is gener-ally small. Consequently temperature has a relatively small influence on the density and there is little information published on temperature fluctuations in estuaries. One can visualize estuaries, however, where temperature could be a dominant factor at times. Many tropical estuaries have little river flow entering them during the hot season. Surface heating could then provide sufficient density difference between the estuary and the sea to maintain a gravitational circulation. Because of the diurnal variation of temperature, however, these effects would be transitory. In many fjords there is no river discharge in winter and surface cooling is intense. The surface waters can then become more dense than those at depth and will tend to sink. This vertical circulation phenomenon is known as thermohaline convection. The effects of temperature, therefore, must not be forgotten.

Estuaries are formed in the narrow boundary zone between the sea and the land and their life is generally short. Their form and extent is being

1

constantly altered by erosion and deposition of sediment and drastic effects are caused by a small raising or lowering of sea level. These sea level alterations may be eustatic, variations in the volume of water in the oceans, or isostatic, variations in the level of the land. In the recent geological past there have been very large eustatic changes in sea level. About 18000 years ago the sea level stood about 100 m below its present level, the water being locked up in extensive continental ice sheets. As the ice retreated the sea rose at a rate of about 1 m a century, drowning the valleys incised by the rivers. This Flandrian transgression ended about 3000 B.C. when sea level was more or less the same as at present. Since then some authorities have suggested that minor fluctuations have occurred, but these are probably mainly isostatic in origin. Scotland is rising at a rate of about 3 mm yr^{-1} in response to the removal of the ice sheet, whereas areas formerly peripheral to the ice sheet, such as southern England and Holland, are now sinking at about 2 mm yr^{-1}.

Further large eustatic changes are possible. If all the world's ice were melted, it has been estimated that sea level would rise by about 30 m. If this happened, new estuaries would be formed in the upper valleys of the present rivers. Little sediment would appear from the rivers, but large quantities would be available from renewed coastal erosion. A reduction of sea level would produce shallow estuaries which would quickly fill with sediment derived from the rejuvenated upper river valleys. In either case, because of the increased, or reduced, depths of the gulfs and seas into which the estuaries emptied, the tidal conditions would be modified. At present, following the Pleistocene ice age which overdeepened the river valleys, and the subsequent inundation which flooded them, estuaries are both well-developed and numerous. In geological terms this situation may not last long.

Though they are a particularly ephemeral feature of the earth's surface, estuaries have probably been extremely important in the world's development. They have generally high inflows of nutrients from the land, but, because of their range of conditions, tend to have a lesser diversity of life than other aquatic environments. Individual species are numerous, but are specialized and often tolerant to large extremes of temperature and salinity. It is thought by many zoologists that estuaries may have been the most likely situation in which the first signs of organic life evolved. Almost certainly the estuaries were the route by which, many millions of years later, animal life slowly adapted itself to a land-living and air-breathing existence.

Because of their fertile waters, sheltered anchorages and the navigational access they provide to a broad hinterland, estuaries have been the main centres of man's development. The promotion of trade and industry has led to large-scale alteration of the natural balance within estuaries by alteration of their topography, making for easier navigation for larger ships, and large-scale pollution, as a result of industrialization and population increases. Deforestation of the land leads to increased run-off from the land, increased flashiness of the discharge and increased sediment load in the rivers. Building and paving of large areas also produce a quick response of run-off to rainfall. These effects may be controlled by the building of dams and may be reduced by the removal of river water for industrial processes and household use. However, maintenance of river flow at a set level will decrease the natural tendency for rivers to flush sediments out of their estuaries and consequently may aggravate shoaling problems. Deepening of the estuary by dredging will increase the estuary volume and reclamation of intertidal areas will decrease the tidal flow, alter the mixing processes and circulation patterns and perhaps decrease the flushing time of the estuary. With decreased flushing time the estuary cannot cope with and dispose of such large quantities of effluent. To understand and to be able to predict these effects is essential if mankind is not to do undue damage to his environment.

The main drawback in studying estuaries is that river flow, tidal range and sediment distribution are continually changing and consequently some estuaries may never really be steady-state systems, they may be trying to reach a balance they never achieve. Because of the interaction of so many variables no two estuaries are alike and one never knows whether one is observing general principles or unique details.

Many studies of individual estuaries are available in the literature and there are several texts that sift the details and produce the relevant general principles. These include Cameron and Pritchard (1963), Lauff (1967) and Ippen (1966). In this book some of the many examples will be discussed and used as illustrations of the techniques of analysis and computation useful for estuarine studies.

CHAPTER 2

Classification of Estuaries

In order to compare different estuaries and to set up a framework of general principles, within which it may be possible to attempt prediction of the characteristics of estuaries, a scheme of classification is required. Many different schemes are possible, depending on which criteria are used. Topography, river flow and tidal action must be important factors that influence the rate and extent of the mixing of the salt and fresh water. Locally, and for short periods, wind also may become significant. The resultant mixing will be reflected in the density structure and the presence of stratification may cause modification of the circulation of water. Obviously all of these causes and effects are interlinked and it would be difficult to take account of them all in one classification system.

CLASSIFICATION BY TOPOGRAPHY

A topographic classification has been presented by Pritchard (1952). He divides estuaries into three groups: coastal plain estuaries, fjords and bar-built estuaries.

Drowned River Valleys (coastal plain estuaries)

These estuaries were formed during the Flandrian transgression by the flooding of previously incised valleys. Sedimentation has not kept pace with the inundation and the estuarine topography is still very much like that of a river valley. Consequently maximum depths in these estuaries are seldom as much as 30 m. They have the cross-section of subaerial valleys and deepen and widen towards their mouths, which may be modified by spits. Their outline and cross-section are both often triangular. The width–depth ratio is usually large, though this depends on the type of rock in which the valley was cut. Extensive mudflats and saltings often occur and the central channel is often sinuous. The entire estuary is usually floored by varying thicknesses of recent sediment, often mud in the upper reaches, but becoming increasingly sandy towards the mouth. A remarkable characteristic of some is that the increase in cross-sectional area towards the mouth is exponential; this may reflect a long-term equilibrium adjustment between sedimentation and erosion by tidal currents.

4

Coastal plain estuaries are generally restricted to temperate latitudes, where, though river flow may be large at times, the amount of sediment discharged by the river is relatively small. River flow is generally small compared with the volume of the tidal prism (the volume between high and low water levels).

Examples: The Chesapeake Bay estuary system in the United States and the Thames, Southampton Water and the Mersey in England.

Fjords

Fjords were formed in areas covered by Pleistocene ice sheets. The pressure of the ice overdeepened and widened the pre-existing river valleys, but left rock bars or sills in places, particularly at the fjord mouths and at the intersection of the fjords. These sills can be very shallow. In Norway a number have sill depths averaging $4\frac{1}{2}$ m and their presence can restrict the free exchange of water with the sea. The inlets of British Columbia, however, have deeper sills. Pickard (1956) has listed the physical features of several of these inlets and the sill depths are mainly between about 40 m and 150 m. Inside the sills the maximum depth of the inlets reaches almost 800 m.

Because of overdeepening, fjords have a small width–depth ratio, steep sides and an almost rectangular cross-section. Their outline is also rectangular, but sharp, right-angled bends are common. Some fjords reach 100 km in length and the width–depth ratio is commonly 10 : 1.

Fjords generally have rocky floors, or very thin veneers of sediment, and deposition is generally restricted to the head of the fjord where the main rivers enter. River discharge is small compared with the total fjord volume, but, as many fjords have restricted tidal ranges within their mouths, the river flow is often large with respect to the tidal prism. Their occurrence is restricted to high latitudes in mountainous areas.

Examples: Loch Etive (Scotland), Sogne Fjord (Norway), Alberni Inlet (British Columbia) and Milford Sound (New Zealand).

Bar-built Estuaries

These estuaries could also be called drowned river valleys as they have experienced incision during the ice age and subsequent inundation. However, recent sedimentation has kept pace with the inundation and they have a characteristic bar across their mouths. This bar is normally the break-point bar formed where the waves break on the beach and for this to be well developed the tidal range must be restricted and large volumes of sediment available. Consequently, bar-built estuaries are generally associated with depositional coasts. The estuaries are generally only a few metres deep and often have extensive lagoons and shallow waterways just inside the mouth. Because of the restricted cross-sectional area current velocities can be high at the mouth, but in the wider parts further inland they rapidly diminish.

The river flow is large and seasonally variable and large volumes of sediment are riverborne at times of flood. The estuary form is governed by the river regime at the flood stage and may show a basin–bar structure caused by meander scouring. During the floods the bar may be swept completely away, but will quickly re-establish itself when the river flow diminishes. The mouth may undergo considerable variations in position from year to year. Variations of up to 3 km have been recorded for the mouth of the Vellar and the coast-line around the estuary has built out by 300 m between 1931–1967.

Bar-built estuaries are generally found in tropical areas or in areas with active coastal deposition of sediments.

Examples: Vellar Estuary (India), Roanoke river (United States).

The Rest

In this section one can include all estuaries that do not conveniently fit elsewhere. Included are tectonically produced estuaries: estuaries formed by faulting, landslides and volcanic eruptions.

Examples: San Francisco Bay, where the lower reaches of the Sacramento and San Joaquin Rivers have been drowned by movements on faults of the San Andreas fault system.

CLASSIFICATION ON SALINITY STRUCTURE

The majority of estuaries that have been studied fall within the coastal plain category and it is apparent that within this group large differences occur in the circulation patterns, density stratification and mixing processes. Consequently, a better classification would be one based on the salinity distribution and flow characteristics within the estuary. Examination of such a classification will lead to a better understanding of how the circulation of water in the estuaries is maintained.

Pritchard (1955) and Cameron and Pritchard (1963) have classified estuaries by their stratification and the characteristics of their salinity distributions. They define four main estuarine types: highly stratified or salt wedge, fjords, partially mixed and homogeneous. The last group is subdivided into laterally inhomogeneous and sectionally homogeneous.

The Highly Stratified Estuary. Salt Wedge Type

Let us first consider an estuary emptying into a tideless sea, with a source of fresh water at its upper end. Also let us consider that the situation is frictionless, that the water behaves as a fluid without viscosity. Under these conditions the river water, being less dense than the sea water, flows outwards over the surface of the saline layer. The velocity in the surface layer decreases towards the mouth as the estuary widens. The interface between the fresh and salt water would be horizontal and would extend a distance up the estuary as far as mean sea level. Because of Coriolis force the seaward flowing river water would be concentrated on the right-hand side (looking downstream) in the northern hemisphere. There would be no mixing of salt and fresh water and no motion at all in the saline wedge. The velocity and salinity profiles would be as shown in Figure 2.1. The velocity would become zero at the upper surface of the salt wedge as defined by the salinity profiles, and would decrease towards the mouth as the estuary widens.

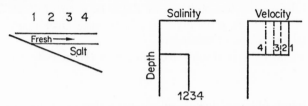

Figure 2.1   Salinity and velocity profiles in a frictionless estuary

Let us introduce friction in the form of viscosity. There will now be a shear in the fluid flow near the interface and the salt wedge will be pushed downstream until its upper surface has a slope sufficient to resist this force. The tip of the salt wedge will become blunted and the water surface will slope more steeply towards the sea. Coriolis force will now affect the lateral water slopes, with the interface sloping downwards toward the right and the sea surface downwards towards the left in the northern hemisphere.

Because of the velocity shear across the interface, a thin layer at the top surface of the salt wedge will be swept seawards. When the shear is sufficiently intense waves form and break on the interface and salty water is mixed into the surface fresh water. This process is called 'entrainment' (see chapter 3), and it is strictly a one-way process. In order to preserve continuity a slight compensating landward flow is necessary in the salt wedge, to replace the salt water passing into the upper layer. The entrainment adds volume to the water flow in the surface layer in its passage down the estuary and conse-quently the discharge increases towards the mouth. The salinity in the wedge will be virtually constant along the estuary. Typical velocity and salinity profiles are shown in Figure 2.2. It is noticeable now that the velocity falls to zero below the upper surface of the salt wedge, as defined by the maximum salinity gradient.

For this type of estuary, the ratio of river flow to tidal flow must be large and generally the ratio of width to depth is relatively small. The mouth of the Mississippi has a diurnal tide with a mean range of about 70 cm and the river discharge is between $8 \cdot 5 \times 10^5$ and $2 \cdot 8 \times 10^5$ m^3 sec^{-1}. In the Southwest Pass upstream flow prevails in the salt wedge regardless of the tidal phase, with downstream flow on the surface. This Pass is dredged to maintain adequate navigational access. In the shallower South Pass upstream flow occurs in the

wedge during the flood tide at the same time as downstream flow occurs on the surface. During the ebb tide the flow in the wedge can be reversed and the current can be in a seaward direction at all depths (Wright, 1971).
Examples: Mississippi and Vellar estuaries.

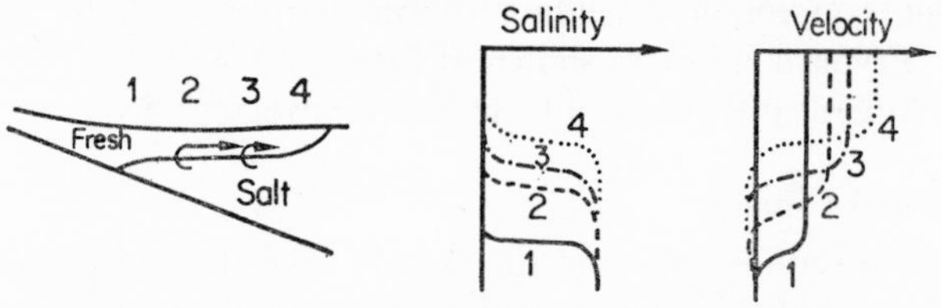

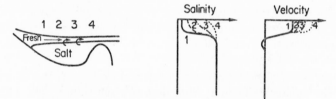

Figure 2.2   Salinity and velocity profiles in a salt wedge estuary

Highly Stratified Estuary. Fjord Type

In many ways these estuaries are similar to the salt wedge type. The lower, almost isohaline layer, is, however, very deep. As river flow is dominant over tidal flow, entrainment is again the process mixing the fresh and salt waters. The upper layer is commonly of virtually constant thickness from head to mouth, but discharge again increases towards the mouth. In some fjords the upper layer thickness is restricted to a depth equal to the sill depth.

Where river discharge is high, the surface layer is almost homogeneous and the maximum salinity gradient occurs below the surface. Where run-off is lower, and near the fjord mouths, the surface layer is less homogeneous and the maximum salinity gradient occurs at the surface (Pickard, 1961). Though

Figure 2.3   Salinity and velocity profiles in a fjord

the temperature generally decreases with increasing depth, there are many instances where there are several marked maxima and minima. These occur especially in fjords with inflow of melt water from glaciers (Pickard, 1971).

Because of the larger tidal velocities and weaker stratification, the circulation over the sills may be entirely different from that occurring within the fjord. Generally, the inflow over the sill is composed of a mixture of the coastal water and the outflow water. In the deeper parts of the fjord tidal action is small and there is often a layered structure showing successive intrusions of saline water. Often this renewal occurs annually and occasionally, where the sill depths are small, renewal is so infrequent that anoxic conditions develop near the bottom. The saline inflow is best developed in the summer when the river flow is largest, entrainment is most active and the density difference between the deep fjord water and the coastal water is greatest. In winter, when river discharge is low, surface cooling can produce thermohaline convection extending to the bottom. Salinity and velocity profiles for a typical fjord are shown in Figure 2.3.

Examples: Alberni Inlet (British Columbia), Silver Bay (Alaska).

Partially Mixed Estuary

If we now introduce tides into the estuary then the entire contents of the estuary will oscillate. It requires only a small tidal range to make this occur, though there is likely to be a particular range of tidal prism to estuary volume ratio over which it is effective in producing a partially mixed estuary.

The energy involved in these movements is large and it is mainly dissipated in the estuary by working against the frictional forces on the bottom, producing turbulence (see chapter 3). The turbulent eddies lose their kinetic energy by working against the density gradients, thereby increasing the potential energy of the water column, and by viscous dissipation, creating heat. These eddies can mix both salt water upwards and fresh water downwards. Consequently the salinity of the surface layer is considerably raised and in order to discharge a volume of fresh water equal to the river flow the seaward surface flow is enhanced. This causes an increase in the volume of the compensating landward flow. Consequently a distinct two-layer flow system is developed. Pritchard has calculated that in the James estuary the seaward flow in the upper layer is twenty times the river flow and the compensating inflow on the bottom is nineteen times the river flow. It is now more difficult, however, to measure these flows because they are small relative to the oscillatory tidal flow superimposed on them. Some sort of averaging process is necessary to determine them (see chapter 3).

Because of the efficient exchange of salt and fresh water the salinity structure of the estuary is different from that of a salt wedge type. The surface salinity increases much more steadily down the estuary and undiluted fresh water only occurs very near the head of the estuary. Also in the saline water on the bottom there is a longitudinal gradient of salinity. Consequently there is a large section in the middle reaches of this type of estuary in which the horizontal salinity gradients are almost linear. The form of the vertical salinity profile does not change much along the estuary either. There is normally a zone of high salinity gradient at about mid-depth and the surface and bottom layers are almost homogeneous. In the shallower parts on the estuary sides, however, the homogeneous bottom layer is missing and the maximum salinity gradient may occur near the bottom. In the upper layer surfaces with equal pressures slope towards the sea and in the bottom layer they slope towards the land. The lateral slopes are affected by Coriolis force to a marked degree and are intensified as both outflowing and inflowing waters are deflected to opposite sides of the estuary.

The river flow must now be low compared with the tidal prism. In Southampton Water, for instance, the maximum river flow is about 50 m³ sec⁻¹, the maximum tidal flow is 7500 m³ sec⁻¹ and the tidal prism is about 10⁸ m³. Typical salinity and velocity profiles are shown in Figure 2.4.

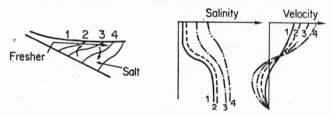

Figure 2.4   Salinity and velocity profiles in a partially mixed estuary

The relationship between tidal amplitude, tidal currents and salinity in this type of estuary is interesting. If the estuary is of the correct depth and length then it is possible for the tidal wave to enter, be reflected from the upstream end and return in a time equal to a harmonic of the tidal period. The reflected wave will then interfere with the wave just entering. A standing wave system can thus be set up in the estuary. Commonly the node is produced near the estuary mouth with the antinode at the head, but in longer estuaries several nodes and antinodes can be present. With a single node at the mouth the tidal amplitude increases toward the head, but the maximum currents occur near the mouth. High and low waters and the time of turn of the current are simultaneous throughout the estuary. The tidal amplitude and salinity variation are thus 90° out of phase with the current velocity (Figure 2.5a).

If the energy of the tidal wave is completely dissipated before reflexion then the tidal wave becomes solely progressive in nature. The amplitude of the tide and the magnitude of the tidal currents diminish towards the head of the estuary and there is a progression in the times of high and low water and turn of the current along the estuary. In this case the tidal amplitude and the current velocity would be in phase, i.e. maximum flood currents would occur at high water (Figure 2.5b). Though the tide curve may be symmetrical outside the estuary, it tends to become asymmetrical inside. As each part of the tidal wave travels at a speed which depends on the water depth, the crest travels faster than the trough. The water level consequently shows a quicker rise and a slower fall. As an estuary shallows and narrows towards the head, the tidal amplitude will tend to increase upstream because of the convergence, but decrease because of friction. Amplification, especially where the tidal range is normally large, can make the tidal amplitude no longer small

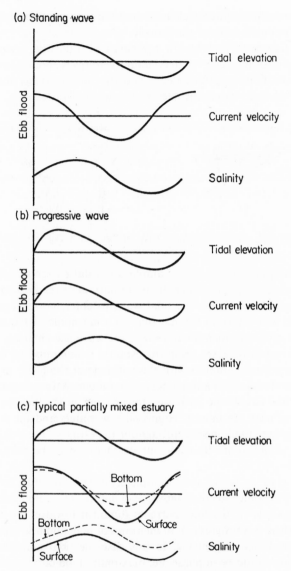

(a) Standing wave

Tidal elevation

Current velocity

Salinity

(b) Progressive wave

Tidal elevation

Current velocity

Salinity

(c) Typical partially mixed estuary

Tidal elevation

Bottom

Current velocity

Bottom

Surface

Salinity

Surface

Figure 2.5 The tidal response in estuaries

compared with the depth, the asymmetry can become very marked and a 'bore' can develop. In long shallow estuaries where the velocity of propagation of the tidal wave is small, several progressive tidal waves can be present at the same time.

As there is some dissipation of the tidal energy before and after reflexion, the tidal response in most estuaries is a mixture of a standing wave with a progressive contribution of variable magnitude. Depending on the relative magnitudes of the two influences so the tidal amplitude and the timing of events vary along the estuary. At Cuxhaven on the Elbe, for instance, the flood and ebb currents start about $1\frac{1}{2}$ hours after low and high waters. The time of high water propagates up the estuary at about 6 m sec^{-1}, and the tidal amplitude diminishes gradually towards the head of the estuary. Though the lag of the turn of the current after low water seems to be about the same all the way up the estuary, the flood current alternates closer to high water further upstream (Defant, 1961). When an estuary is partially mixed and the river flow is mainly confined to the surface layer, the ebb flow will commence on the surface at the head of the estuary as soon as the pressure of the incoming high water diminishes and the turn will progress downstream. Similarly, at the mouth the outflowing surface water will not be stemmed until there is a pressure gradient opposing it and the turn of the current then progresses upstream. Consequently, the bottom current is landward at the beginning of the flood before the surface changes, and the surface current ebbs before the bottom current. At a station in the middle estuary, these effects produce current velocity variations similar to that shown in Figure 2.5c. There is a significant difference in the amplitudes of the surface and bottom tidal velocity fluctuations and the average current over a complete tidal period is in the flood direction at the bottom and in the ebb direction at the surface. The landward residual flow in the bottom layer obviously diminishes towards the head of the salt intrusion. At that point there will be a null point landwards of which, in the inner part, the bottom velocities will be downstream. At this null point shoaling tends to occur and there is a zone of maximum turbidity caused by suspended material trapped in the water circulation (this is further discussed by Dyer, 1972).

Examples of partially mixed estuaries include the James River, the Mersey and Southampton Water.

The Vertically Homogeneous Estuary

When the estuary cross-section is small the velocity shear on the bottom may be large enough to mix the water column completely and make the estuary vertically homogeneous. It is difficult to be sure, however, that vertically homogeneous estuaries really exist, as small vertical variations may be lost in the averaging processes. When there is no vertical salinity gradient there is no vertical flux of salt and mixing occurs only in the horizontal direction. In these estuaries tidal flow will be much larger than river flow.

(a) Laterally inhomogeneous

When the estuary is sufficiently wide Coriolis force will cause a horizontal separation of the flow. The seaward net flow will occur at all depths on the right-hand side in the northern hemisphere and the compensating landward flow on the left. Thus the circulation would be in a horizontal plane rather than in the vertical sense as found in the other estuarine types. The increase of salinity towards the mouth will be regular on both sides of the estuary (Figure 2.6). The lower reaches of the Delaware and Raritan estuaries are examples of this type.

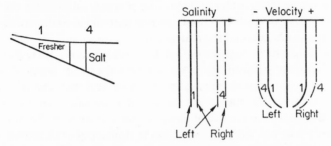

Figure 2.6 Salinity and velocity profiles in an homogenous estuary with lateral variation in the Northern Hemisphere

(*b*) *Laterally homogeneous* (sectionally homogeneous)

When the width is smaller, lateral shear may be sufficiently intense to create laterally homogeneous conditions. Salinity increases evenly towards the mouth and the mean flow is seawards throughout the cross-sections. This flow would tend to drive the salt out of the estuary. The balance is made by an upstream turbulent exchange of salt that is associated with the effect on the tidal flow of topographic irregularities and friction at the bottom. Salty water is trapped in embayments at high water and slowly bleeds back into the main body of water during the ebb stage. It is unlikely that this would be an effective mixing process unless the longitudinal salinity gradients are large. Consequently, the length of the estuary in which salinity is measurable may only be the length of a few tidal excursions. In this type of estuary the effect of friction on the tidal wave is likely to be larger than in other estuarine types, and the tidal wave is likely to have a larger progressive component. When this occurs the maximum flood current is near high water, when the cross-sectional area is large, and the maximum ebb occurs near low water when the cross-sectional area is smaller. As a result there is a larger mass transport of salt on the flood than on the ebb and this mass transport on the progressive tidal wave can help balance the seaward transport on the mean flow.

There is obviously some correspondence between the results of classification on topographical and salinity structure bases. It is clear though, that the limits of each estuarine type are never well defined. The different types are merely stages on a continuous sequence. This sequence will obviously be dominated to a certain extent by the ratio of river flow to tidal flow. Simmons

(1955) has found that when the flow ratio (the ratio of river flow per tidal cycle to the tidal prism) is 1·0 or greater, the estuary is highly stratified. When the flow ratio is about 0·25 the estuary is partially mixed and less than 0·1 well mixed.

This is a very general statement of the situation, as estuary width and depth will have some control on the amount of mixing that a particular tidal rise and fall will produce in the water column. The Mersey and Southampton Water have flow ratios of 0·01–0·02 and yet are partially mixed, having vertical salinity differences commonly greater than 1 %.

The estuarine type may show variation from section to section of the estuary. Near the head of the estuary where the tidal amplitude may be reduced, river flow can dominate, entrainment be active and a salt wedge structure will result. Further downstream tidal velocities can increase, eddy diffusion can become more active and a partially mixed structure will occur. Of course the flow ratio may well change drastically with season and the estuarine type will change in sympathy.

Topographic differences are important factors influencing the flow structure in an estuary. Pritchard (1955) considers estuary depth and width to be important parameters controlling an estuary's position in the sequence. If the river flow and the tidal range are kept constant and the estuary width is increased, the ratio of tidal volume to river flow is changed, acting similarly to a decrease in river flow. This will tend to cause a more completely mixed estuary. Similarly, increasing the depth will decrease the ratio of river flow to tidal flow, but the effect of this will be offset by decreasing the effectiveness of the vertical tidal mixing. Thus the river flow becomes confined to the surface and the estuary becomes more stratified. These effects are quite often shown near constrictions in estuaries. The constricted sections where tidal current velocities are greater tend to be well mixed, whereas the wider pools are more stratified. Where the Hudson River empties into New York Harbour, the abrupt widening causes the structure to change from well mixed to partially mixed. At this point, the landward bottom flow meets the seaward flow in the well mixed part and shoaling problems occur (Duke, 1961).

CLASSIFICATION USING A STRATIFICATION PARAMETER

As already indicated, the amplitude and phase of the tidal wave varies along the estuary. A proportion of the energy lost by the tide is used in mixing, by increasing the potential energy of the water column. Ippen and Harleman (1961) have developed expressions that relate the relative times of high water along the estuary σ_{tH}, and the tidal amplitudes, to the phase change kx (the wave number $k = 2\pi/\lambda$, λ being the wavelength of the tidal wave), and a damping coefficient μ which specifies the change in amplitude with distance x along the estuary caused by friction.

The time of high water at any position in the estuary relative to high water at the estuary head σ_{tH} is given by:

$$\tan \sigma_{tH} = -\tan kx \tanh \mu x$$

In a channel of uniform cross-section and roughness, k and μ would be expected to be constants and

$$\mu = \frac{\phi}{2\pi} k$$

where ϕ is a dissipation constant. The values of ϕ, μ and k can be determined for any estuary by using the nomogram Figure 2.7, with measured values of

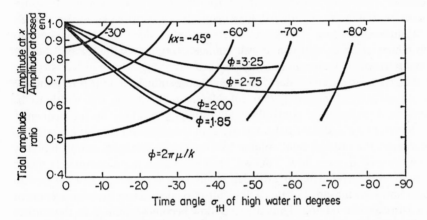

Figure 2.7   Nomogram for determination of μ and k from tidal amplitudes and time of high water. (Reproduced with permission from A. T. Ippen and D. R. F. Harleman, 1961, *Tech. Bull. 5*, Comm. Tidal Hyd. Corps. Engineers U.S. Army, Figure 6)

the time angle of high water and the relative tidal amplitudes measured at a number of points at distances x from the head of the estuary.

The rate of transport of tidal energy (Px) across any section is

$$Px = -cbg\rho A_0^2 \sinh 2\mu x$$

where the wave velocity $c = 2\pi/Tk$, T being the tidal period, b is the breadth, ρ is the water density and A_0 the tidal amplitude at the estuary head. For a purely standing wave without any progressive component $\mu = 0$ and the total energy flux is zero.

The rate of energy dissipation in the portion of a channel between two cross-sections x_1 and x_2 is $Px_1 - Px_2$ and the rate of energy dissipation per unit mass of water is

$$G = (Px_1 - Px_2)/\rho bh(x_1 - x_2)$$

As a water particle moves down the estuary towards the sea, it gains potential energy due to its increasing density. The rate of gain of potential energy per unit mass over the entire length of the estuary L is

$$J = g\left(\frac{\Delta\rho}{\rho}\right)hu_f/L$$

where $\Delta\rho$ is the density difference between the fresh and ocean water and u_f is the mean velocity of fresh water over the distance L. Thus for a given estuary J is affected only by variations in river discharge and G indicates the amount of energy dissipated from the tide that can either mix the water column or be liberated as heat. The ratio G/J, the stratification number, is thus a measure of the amount of energy lost by the tidal wave relative to that used in mixing the water column. This ratio is analogous to the inverse of a Richardson number (chapter 3). Increasing values of the stratification number indicate increasingly well mixed conditions, and low numbers, highly stratified conditions (Figure 2.8). If other factors are maintained constant, increasing river flow reduces the stratification number, indicating

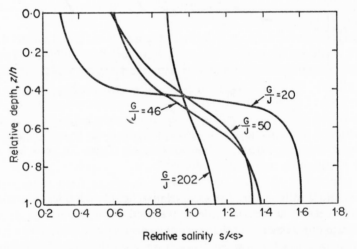

Figure 2.8   Vertical salinity gradients in relation to stratification number, derived from model experiments. (Reproduced with permission from A. T. Ippen and D. R. F. Harleman, 1961, *Tech. Bull. 5*, Comm. Tidal Hyd. Corps Engineers, U.S. Army, Figure 20)

increased stratification. However, as the stratification number is also dependant on the breadth and depth of the estuary, similar river discharges and tidal conditions in estuaries of different dimensions will produce different stratification numbers. This indicates that the flow ratio does not produce a good quantitative comparison of estuarine stratification.

An example of this method, used successfully in the well mixed coastal plain estuary of Coos Bay, is discussed by Blanton (1969). Its use in the Bay of Fundy and in the Delaware Estuary is discussed in Ippen (1966). This method, however, requires rather precise tidal elevation measurements at several positions in an estuary and knowledge of the river flow.

CLASSIFICATION USING A STRATIFICATION–CIRCULATION DIAGRAM

A further quantitative means of classifying and comparing estuaries, and one which requires measurements of salinity and velocity only, has been developed by Hansen and Rattray (1966). A more complete description of their methods of analysis is discussed in chapter 8. They have used two dimensionless parameters to characterize estuaries: a stratification parameter $\delta S/S_o$, defined as the ratio of the surface to bottom difference in salinity (δS) divided by the mean cross-sectional salinity (S_o), and a circulation parameter u_s/u_f, the ratio of the net surface current to the mean cross-sectional velocity. The circulation parameter expresses the ratio between a measure of the mean fresh water flow plus the flow of water mixed into it by entrainment or eddy-diffusion, to the river flow.

Their classification diagram is shown in Figure 2.9. In Type 1 the net flow is seaward at all depths and upstream salt transport is by diffusion. Type 1a has slight stratification and coincides with the laterally homogeneous well mixed estuary. In Type 1b there is appreciable stratification. In Type 2 the flow reverses at depth and corresponds to the partially mixed estuary. Both advection and diffusion contribute to the upstream salt flux. In Type 3 the salt transfer is primarily advective: in Type 3b the lower layer is so deep that circulation does not extend to the bottom, e.g. fjords. Type 4 has more intense stratification, salt wedge type.

The uppermost boundary represents conditions of fresh water outflow over a stagnant saline layer. The separation of the classes is again somewhat arbitrary, and the data points from several estuaries show that estuaries are characterized by a line rather than a point. This arises because of two related effects; changes in river flow can cause a point on the estuary to change its position on the diagram, change its type even, and different sections in the estuary can have different positions on the diagram. Consequently, change in river flow is equivalent in a general way to change in position along the estuary. As river flow increases there is a tendency for the salinity and mixing structure to be forced further down the estuary and vice versa. This classification system is probably the most satisfactory one at present yet proposed.

There are also, of course, exceptions showing different characteristics to the generalized estuarine sequence. One well documented example is that of Baltimore Harbour, a small tributary of the Chesapeake Bay. There is very

2

little river flow into the Harbour to contribute to the density stratification, but in Chesapeake Bay there is a normal stratification. Within the Harbour tidal mixing processes are intense and the surface salinity is higher than that in the Bay and the bottom salinity is lower. At mid-depth the salinities are similar. This leads to a three-layer flow system with inflow of fresher water on the surface and of salty water on the bottom and outflow of intermediate density at mid-depth.

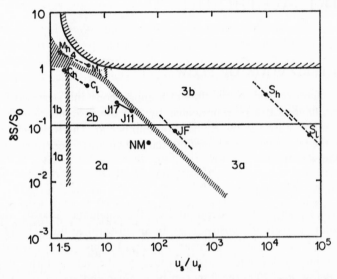

Figure 2.9 Classification diagram with some examples. (Station code: M, Mississippi River mouth; C, Columbia River estuary; J, James River estuary; NM, Narrows of the Mersey estuary; JF, Strait of Juan de Fuca; S, Silver Bay. Subscripts h and l refer to high and low river discharge; numbers indicate distance (in miles) from mouth of the James River estuary. (Reproduced with permission from D. V. Hansen and M. Rattray Jr., 1966, *Limnol. Oceanog.*, **11**, 319, Figure 2)

Similar three-layer flow effects have been found when strong winds blow up estuaries. The surface flow becomes reversed over the top few metres. Below this the outflow is considerably enhanced. The increased mixing tends to decrease the normal salinity stratification, but the normal pattern becomes re-established within a few days of the wind action ceasing.

Wind effects are also important in promoting mixing in the shallower estuaries, especially bar-built estuaries. During periods of strong winds the surface wave action can break down the normal stratification.

CHAPTER 3

Entrainment, Turbulence and Averaging

CHARACTERIZATION OF FLOW

Certain characteristics of the flow of homogeneous and stratified fluids in pipes and channels can be represented by two dimensionless numbers, the Reynolds number Re and the Richardson number Ri. The Reynolds number compares the relative importance of inertial and viscous forces in determining the resistance to flow.

$$Re = \frac{uD}{v} \tag{3.1}$$

where u is a velocity, D is a depth and v is the kinematic viscosity, the ratio of molecular viscosity to density (μ/ρ). In an unstratified fluid D will be the total depth of water and u the mean velocity. In this context, whether the flow is laminar or turbulent is determined by the value of the Reynolds number. Below $Re \sim 2000$ the flow can be laminar and above about 10^5 the flow is likely to be fully turbulent. Between these two points the flow is transitional and its character and the point at which it becomes fully turbulent depend largely on the roughness of the walls of the pipe or channel. Sternberg (1968) examined the flow in a number of tidal channels and found that fully turbulent flow occurred at an Re greater than $1 \cdot 5$ to $3 \cdot 6 \times 10^5$ and that the flow over geometrically simple beds became fully turbulent at a lower Re than over beds of complex roughness. In unstratified conditions in rivers and estuaries the flow is always transitional or fully turbulent.

When the fluid is stratified the density gradient resists the exchange of momentum by the turbulence and a velocity shear is necessary to cause mixing. The Richardson number Ri is a comparison of the stabilizing forces of the density stratification to the destabilizing influences of velocity shear, and can be defined by,

$$Ri = -\frac{g}{\rho}\frac{\partial \rho}{\partial z} \bigg/ \left(\frac{\partial u}{\partial z}\right)^2 \tag{3.2}$$

For $Ri > 0$ the stratification is stable, for $Ri = 0$ it is neutral and the fluid unstratified between the two depths, and for $Ri < 0$ it is unstable. When the

19

stratification is above a certain value turbulence will be damped out and the flow will be essentially laminar. This transition from laminar to turbulent flow under conditions of uniform flow is generally taken to occur at $Ri = 0.25$. In estuaries, however, because of non-uniform flow the transition will occur at a higher Ri. Because of the difficulty of precisely measuring the density and velocity gradients and because of the tidal variations, the Richardson number undergoes wide fluctuations in any estuary. Generally, however, conditions are more or less neutral in the surface and bottom layers at times when the tidal currents are near their maximum. The interface, however, is quite likely to be stable. The Columbia River, for instance, has Ri values reaching 5 at mid-depth (Hansen, 1965). At times of less strong tidal currents the whole water column may be stable.

Because of the difficulties of measurement, values of an effective Richardson number have been derived by taking the hourly measurements of $\partial\rho/\partial z$ and $(\partial u/\partial z)^2$ and averaging them over a tidal cycle. This has been done for the Mersey estuary by Bowden (1963) and Bowden and Sharaf el Din (1966a). Near the bottom the effective Ri was of the order 0.1, while at mid-depth and nearer the surface the values were of the order 1.

It is also possible to define a layer Richardson number

$$Ri = \frac{(\Delta\rho/\rho)gD}{u^2} \qquad (3.3)$$

where D is the depth of the upper layer flowing with a velocity u relative to the lower layer and $\Delta\rho$ is the density difference between the layers. In this form the Richardson number is a bulk number reflecting the characteristics of the whole flow rather than the more detailed, localized gradient Richardson number defined by equation (3.2). The square root of the inverse layer Richardson number is the interfacial Froude number F_i.

$$F_i = \frac{u}{\sqrt{(\Delta\rho/\rho)gD}} \qquad (3.4)$$

The Froude number is an alternative way of considering the influence of density stratification and F_i can also be thought of as comparing the velocity of the flow to the velocity of propagation of a progressive wave along a density interface. When F_i approaches unity, the interfacial waves cannot propagate upstream, but they increase in amplitude to such an extent that vigorous vertical mixing ensues.

In estuaries it is sometimes useful to write the velocity in equations (3.3) and (3.4) in terms of the cross-sectional mean velocity associated with the river discharge. The densimetric Froude number then is:

$$F_m = \frac{u_f}{\sqrt{(\Delta\rho/\rho)gh}} \qquad (3.5)$$

where $u_f = R/A$, R being the river discharge, A the cross-sectional area and h the estuary depth.

Also Fischer (1972) has defined an estuarine Richardson number Ri_E. The input of buoyancy per unit width due to the river flow is

$$g\frac{\Delta\rho}{\rho}\frac{u_f}{b}$$

which is then compared with the tidal velocities by writing

$$Ri_E = \frac{g(\Delta\rho/\rho)(u_f/b)}{u_t^3} \tag{3.6}$$

where u_t is the root mean square tidal velocity.

As the density stratification restricts the amount of vertical mixing that can take place, vertical mixing will decrease with decreasing F_m, i.e. with increasing Ri. Thus, as Hansen and Rattray (1966) have shown, F_m is an important parameter reflecting different estuarine types.

MIXING IN STRATIFIED FLOWS

The mixing of salt and fresh water in estuaries is carried out by two processes; entrainment and diffusion. Entrainment is a one-way process in which a less turbulent water mass becomes drawn into a more turbulent layer. The rate of entrainment will increase with increasing velocity difference between the layers, i.e. with increasing F_i. Because of the vertical movement of salt in the entrained water, the potential energy of the water column is increased. As a consequence of entrainment of extra volume into the upper more turbulent layer, the discharge of the more turbulent layer increases downstream. Diffusion, however, is a two-way process in which equal volumes of water are exchanged between the two layers. This requires the presence of turbulence in both layers. Though there is no net exchange of water, salt is transported upwards and the potential energy of the water column is again increased.

The rate of mixing by the two different methods depends on the degree of turbulence in the two layers. If the turbulence is the same in both upper and lower layers there is no entrainment, all the mixing being by turbulent diffusion. If the lower layer is static, then there is no diffusion across the interface and the mixing is entirely by entrainment. Consequently both methods of mixing can work at the same time, their relative importances depending largely on the degree of turbulence in the lower, saline layer.

As estuaries are mainly stratified and turbulent, mixing will be effected by both entrainment and diffusion. To a first approximation the ratio of their contributions will vary with the ratio of river discharge to tidal prism volume.

Carstens (1970), twice using the principles of salt and water continuity on two-layer flow in a Norwegian fjord, found 10% and 100% entrainment. The estimated volume of the entrained water varied seasonally with the river discharge and was between 40–150% of the river discharge.

Entrainment

It has been observed in model experiments that, when a light surface layer is passed over a static, denser, lower layer, for small current shear the density interface is smooth. As the current in the upper layer, and consequently the shear across the interface, is increased, the interface becomes disturbed by waves. These internal waves are three-dimensional, having crest lengths the same order as the wavelengths and small compared with the channel width (Lofquist, 1960). As the shear is further increased, the waves become higher and sharper crested. Eventually they break and elements of the denser water are ejected from the crests into the lighter layer above. This process of breaking internal waves causes entrainment and the limiting velocity above which this process occurs must obviously be related to the degree of density stratification. Little passage of salt into the upper layer occurs until the waves break, then the amount of entrainment will rise as F_i is increased.

Most of the information on entrainment has been derived from laboratory experiments which have attempted to assess the effect on the rate of entrainment of the degree of stratification defined by the Froude number and the intensity of turbulence in the moving layer given by the Reynolds number. From experiments with fresh water and sugar solutions, where a light surface layer with a velocity u was passed over a stationary lower denser layer, Keulegan (1949) has shown that a non-dimensional criterion of mixing Θ could be defined as:

$$\Theta^3 = v_2 g \frac{\Delta\rho}{\rho_1} \cdot \frac{1}{u^3} \tag{3.7}$$

where ρ_1 is the density of the upper layer and v_2 the kinematic viscosity of the lower layer. With an $Re > 450$, the internal waves on the interface broke when the upper layer had a velocity u_c given by the value of $\Theta = 0.178$.

The wavelength λ of the waves at the critical velocity was given by the formula

$$\lambda = \frac{\pi u_c^2}{g \Delta\rho/\rho_1} \tag{3.8}$$

Though the breaking of each wave is a discrete process, it is so continuous in space and time that, when averaged, it can be likened to vertical flow, a velocity of entrainment W_e. Keulegan also showed that

$$\frac{W_e}{(u - 1.15 u_c)} = K \tag{3.9}$$

The factor K varied with $\Delta\rho/\rho_1$ and consequently with F_i, between the limits 2.2×10^{-4} and 3.5×10^{-4} at $\Delta\rho/\rho_1$ of 0.16 and zero respectively.

Keulegan advised against using these results on large bodies of water as his experiments were only for an initial reach where equilibrium was probably not achieved. However, using the criterion for fresh and salt water, u_c would be of the order of a few centimetres per second. In natural circumstances this would mean that internal waves on an estuarine interface would virtually always be breaking.

Lofquist (1960) extended Keulegan's work, using a dense liquid moving beneath a stationary upper layer. He found that there was no consistent dependence of W_e/u on Reynolds number above $Re = 450$. The internal Froude numbers for his experiments were between about 0·02 and 0·55 and W_e/u increased rapidly with increasing F_i from about 3×10^{-5} for $F_i = 0·1$ to about 6×10^{-4} at $F_i = 0·6$. This data can be made compatible with that of Keulegan by using the same origin for measuring the length scales. Macagno and Rouse (1962) have studied mixing between two layers moving in opposite directions and found that W_e/u varied with both Froude and Reynolds numbers. This, however, was a situation where eddy–diffusion may also have been present.

There is considerable difficulty in comparing these and other experimental results because of the various experimental conditions and the slightly differing definitions of the length involved in the Reynolds and Froude numbers. Also, because of problems of scaling, these results involving entrainment cannot with certainty be applied to natural estuaries. However, similar processes are probably operating naturally. A vivid description of mixing in a natural estuary has been given by Farrell (1970) describing the Saco River in Maine; 'Ninety minutes after high water, a strong fresh-water current, with a surface flow of up to 5 feet per second, had developed, while less than 6 feet down current velocities at five stations were less than 1·0 feet per second. The effect was most pronounced at Station 9 where the velocity of 5 feet per second existed only in the top 3 feet of the water column. So pronounced was this fresh-water–salt-water boundary that a diver approaching it from below in the virtually still seawater could observe the brown river water flowing overhead in violent turbulence. The actual boundary was so well defined that it was limited to a 6-inch width. Occasionally, vortices distorted the boundary to a depth of 4 or 5 feet below the boundary surface. This caused waveform undulations to spread upstream until the vortex subsided and the straighter plane of shearing flow re-established itself.'

The Form of the Salt Wedge

Because of the entrainment both salt and volume have been added to the upper layer in its passage down the estuary. Consequently, its discharge increases towards the mouth. The slope of the density interface at the top of the salt wedge will be determined by the magnitude of the frictional stresses on it. These stresses vary with F_i and Re. From continuity considerations

Stommel (1953b) has shown that the depth of the upper layer varies with the transport. Entrainment of salt water from below increases the thickness of the upper layer towards the mouth for F_i less than 0·5. For F_i greater than 0·5 and less than unity the thickness of the fresh surface layer decreases. A similar variation of upper layer thickness has been described by Tully (1949) in Alberni Inlet, with changes in river discharge.

Stommel and Farmer (1952) have shown that the discharge at the mouth of an estuary can be limited by an F_i of unity and there is consequently a critical depth of flow of the upper layer. As the discharge increases, the amount of entrainment increases until at an $F_i \sim 1$ there is a general erosion of the interface causing an increase in the thickness D of the upper layer. This leads to a reduction in the Froude number and maintenance of $F_i \leqslant 1$ for any further increase in velocity. Thus for a given river discharge and density difference the depth of the upper layer can be calculated from $F_i = 1$ in equation (3.4). Similar effects are possible in an estuary with a constriction at the mouth. An $F_i = 1$ appears to occur in the South West Pass of the Mississippi and also has been reported from the South Pass (Wright, 1971). Besides restricting the surface outflow, the limitation of $F_i = 1$ has an effect on the lower layer. Any downwards erosion of the density interface will tend to restrict the compensating inflow occurring in the salt wedge. Because of this, no matter how intense the mixing within the estuary, no increase in the salinity of the outflowing mixture is possible. This 'overmixing' mechanism seems to be an important control on the salinity of harbours such as New York and St. John's, New Brunswick (Stommel and Farmer, 1953). However, the overmixing process is likely to be prevalent mainly in estuaries with constricted mouths, undergoing intense tidal mixing.

The length and shape of the salt wedge in an estuary has been examined by Farmer and Morgan (1953) and Ippen and Harlemen (1961). Farmer and Morgan assumed an estuary without lateral variations, of regular depth and with a rectangular cross-section. They also assumed a sharp density discontinuity, no mixing between the layers and a negligible velocity in the salt wedge. They found that the shape of the wedge was represented by:

$$kF_{i0}^2 \frac{x}{D_0} = \tfrac{1}{6}n^2(3-2n) - F_{i0}^2\left(\frac{n}{1-n} + \log(1-n)\right) \qquad (3.10)$$

where D_0 and F_{i0} are the total water depth and the interfacial Froude number at the tip of the salt wedge, D is the thickness of the wedge at any position x, and $n = D/D_0$, k is the drag coefficient on the wedge. Measurement of D at the mouth, at which $x = L$, for varying values of the mean stream velocity at the wedge tip, enables the drag coefficient to be determined. There was agreement of the theory with laboratory experiments and with the S.W. pass of the Mississippi for which a k of 0·001 was determined. In experimental tests using a k of 0·006, the predicted length of the wedge agreed within 15%

to that observed. The wedge length increased with decreasing discharge and increased with increasing density contrast. A comparison between the predicted and observed wedge shapes is shown in Figure 3.1.

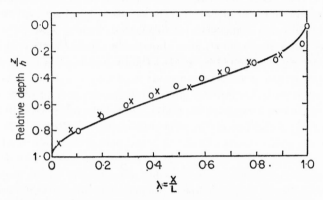

Figure 3.1 Profile of a salt wedge: comparison of results from theory and observations. The solid line is the form of the interface, derived from theory. $\lambda = 0$ at the upstream of the wedge; L is the length of the wedge. O's refer to Mississippi River Southwest Pass and X's refer to a model experiment. (Reproduced with permission from H. G. Farmer and G. W. M. Morgan, 1953, *Proc. 3rd. Conf. Coastal Eng.*, 54–64, Figure 3)

Ippen and Harleman (1961) and Harleman and Ippen (1960) have examined the intrusion length in models of tidal estuaries. In partially mixed estuaries the intrusion length was directly proportional to the square root of the density difference between sea and river waters, inversely proportional to the square root of the fresh water flow rate (u_f) and the sixth root of the rate of energy dissipation per unit mass of fluid in the estuary (G):

$$L \propto \sqrt{\frac{g(\Delta\rho/\rho)}{u_f G^{\frac{1}{3}}}} \qquad (3.11)$$

In well mixed estuaries the intrusion length was inversely proportional to the fresh water flow rate and directly proportional to the cube root of the rate of energy dissipation:

$$L \propto \frac{G^{\frac{1}{3}}}{u_f} \qquad (3.12)$$

Turbulence

It must be very seldom in nature that an estuary can be found discharging into the sea where the influence of tides is absent. Whenever tidal movements are present the velocities are larger than would occur in an ideal salt wedge estuary. The whole mass of water in the estuary is involved in the motion, the Reynolds number is higher and turbulence is more fully developed. The turbulence, as well as mixing salt water into the fresh layer above, also mixes fresh water downwards. Consequently, though there is a net flux of salt there is no net exchange of water and the mixing process is a non-advective one, called eddy-diffusion. The energy for this type of mixing is not derived from the relative motions of two stratified liquids, but from the frictional energy dissipated by the water moving across a rigid bottom.

The turbulent eddies produced in the flow by the current shear against the bottom and sides of the estuary are of fairly consistent form at different positions in the estuary. The large eddies will be anisotropic since they will be limited by the water depth in the vertical scale. Their horizontal scale will be limited by the width of the estuary and the longitudinal scale of the largest eddies will be limited to something like ten times the width of the channel. These eddies are formed continuously and are then passed up or down the estuary on the tidal flow. The actual composition of eddy sizes present in any particular estuary will thus be a complicated combination of the effects of width, depth, current velocity and bottom roughness, and the eddies formed will be modified by the presence of density stratification which will tend to limit their size in the vertical sense.

Consequently, in any estuary there will be a dominant range of eddy sizes, or periods, within which energy is being abstracted from the tidal flow and put into the turbulence. This energy is thought to be dissipated mainly by viscous forces in eddies of very small sizes, which are assumed to be isotropic. Between the large anisotropic eddies and the small isotropic ones the energy is considered to be passed on by a cascade process without significant energy loss. Within this 'inertial sub-range' or 'equilibrium range' the form of the energy spectrum and certain properties of the turbulence can be derived. The spectrum of energy in this range (Figure 3.2) decreases proportional to $k^{-5/3}$, where k is the wave number, which can be defined as $k = 2\pi n/u$, n being the frequency, or $k = 2\pi/uT$, where T is the period.

The frequency or wave number energy spectrum of sea surface waves is well studied, but the energy spectrum of turbulence of tidal currents in estuaries is not so well known. The energy spectrum in a tidal channel in British Columbia was within the inertial sub-range at frequencies greater than 1 Hz (Grant *et al.*, 1962). It is becoming apparent however that the $-5/3$ power law applies in other frequency ranges where there is no energy input. Cannon (1971) has found that in a coastal plain estuary the law is obeyed over large ranges at periods between 1 and 60 minutes. At periods between 1 second and several minutes the spectra are complicated and variable and it appears that most of the energy input is in this range.

Bowden and Proudman (1949) have examined the longitudinal component of turbulence in the Mersey Narrows using a Doodson current meter, both on a stand near the bottom and freely suspended near the sea surface. They found that the turbulent fluctuations were of two types, fluctuations of the order of a few seconds period were superimposed on fluctuations with periods of 30 seconds to several minutes. An arbitrary division at a period of 15 seconds separated the two types. The longer period fluctuations of one period did not usually persist for more than a few oscillations. Intervals of negligible long period variations often occurred, separating sections of record with different predominant periods. The amplitudes of both types of turbulent

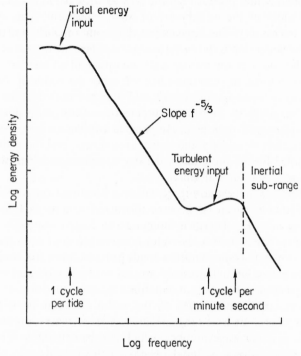

Figure 3.2   Diagrammatic turbulent energy spectrum

fluctuation increased, to a first-order, linearly with mean current and the amplitude of the horizontal fluctuations was about 10% the mean current, i.e. $u'/\langle u \rangle = 0.1$. The amplitudes of the short period fluctuations increased towards the sea bed and were caused by bottom friction. The longer period fluctuations showed no significant variation in amplitude with height above the bottom and resulted either from horizontal eddies or internal waves.

Bowden and Fairbairn (1956) have examined the energy spectrum of horizontal and vertical velocities within 2 m of the sea bed in Red Wharf Bay, off the Isle of Anglesey, using electromagnetic flowmeters. The water column was unstratified. Their measurements were in the frequency range 0·005 to about 0·5 Hz. Both the vertical and longitudinal velocity fluctuations had peak energies within this range. The vertical velocity fluctuations (w′) had a peak energy at a period of 7 sec and the longitudinal velocity fluctuations (u′) a peak energy at a period of 70 sec, but a broader spectrum. As we will see later, cross-products of the form u′w′ are important in the dynamics of estuaries and in this case the major contribution to the mean shearing stress u′w′ would come from fluctuations with frequencies 0·01–0·25 Hz (4–100 sec period).

Further results taken in Red Wharf Bay are presented by Bowden (1962). The spectral peak for u′ was at a wave number $k = 0·25$ m^{-1}, for v′ at $k = 1·5$ m^{-1}, and for w′ at $k = 4·5$ m^{-1}, i.e. 50 sec, 8 sec and 3 sec respectively for u = 50 cm sec^{-1}. Simultaneous measurements showed that the vertical and lateral scales of the u fluctuations were similar and were about $\frac{1}{3}$ of the longitudinal scale, demonstrating the anisotropy of the eddies. The peak contribution to the stress u′w′ was at $k = 0·4$ m^{-1} and most of the shearing stress was due to wave numbers k smaller than 6·0 m^{-1}. The limiting wave number above which the turbulence was isotropic and could not contribute to the stress was given approximately by $kz = 4·0$ where z is the height above the sea bed.

Bowden and Howe (1963) have used the electromagnetic flowmeters near the surface and the bottom in the Mersey and have compared the results with those obtained in Red Wharf Bay. Near the surface in the Mersey a current of at least 100 cm sec^{-1} was needed before turbulence of amplitude greater than 1 cm sec^{-1} was generated, and the main contribution to the energy spectrum of the shearing stress (u′w′) was within the range 0·02–0·15 Hz (50–7 sec period). Near the bottom the energy spectrum of the stress was of higher frequency and was limited by the response of the instruments. The mean amplitudes of the longitudinal fluctuations near the bottom in the Mersey were about half those in Red Wharf Bay and the amplitudes near the surface in the Mersey were about half those near the bottom. The ratio of the w to u fluctuations, however, was very similar in the three situations. The longitudinal scales of the eddies near the sea bed were also very similar, but near the surface they were about three times greater. The differences in intensities of turbulence between the Mersey and Red Wharf Bay may be accounted for by the possible presence of a density stratification in the Mersey.

What happens to the turbulent salinity fluctuation s′ and in particular its cross-products with the velocity fluctuations, e.g. u′s′, is unknown. This is important as it provides the diffusive flux of salt.

AVERAGING

The presence of tidal movements and of turbulence introduce problems that can only be resolved by averaging. One needs to be able to separate the influences of river flow, tidal oscillation and turbulent fluctuations in order to understand the interactions providing conditions of dynamic equilibrium in an estuary.

An instantaneous measurement of velocity u can be represented by three components:

$$u = \bar{u} + U + u' \tag{3.13}$$

where $\bar{u}$ is the mean velocity over a tidal cycle, U is the tidal variation and u' is the short period turbulent contribution. In many estuaries U can be represented moderately realistically by a trigonometrical function such as $U = U_o \sin \omega t$. It is difficult for measuring instruments to record the instantaneous velocity and at once automatic averaging occurs.

Then the observed velocity $u_o = \bar{u} + U$

$$\text{i.e.} \quad u_o = \bar{u} + U_o \sin \omega t \tag{3.14}$$

Taking values over a tidal cycle from $\omega t = 0$ to 2π then

$$\frac{1}{2\pi} \int_0^{2\pi} u_o \, dt = \bar{u} + \frac{1}{2\pi} \int_0^{2\pi} U_o \sin \omega t \, dt = \bar{u}$$

A similar analysis can be carried out for salinity measurements, but with the exception that S is generally about 90° out of phase with velocity and would be represented by $S = S_o \cos \omega t$.

In order to satisfy equation (3.14) a suitable length for each observation must be chosen to separate the essentially harmonic tidal frequencies and the random turbulent ones. To do this the two components must occupy different frequency ranges in the energy spectrum, there must be a gap between the two, and our running averages must be taken over a length of time equivalent to the middle of the gap.

It appears then, from what we have stated about turbulence, that we need to measure over a period of about 100 sec to remove the most significant of the turbulent fluctuations. This agrees with a rough indication which can be achieved by saying that we have to average over a time equivalent to the passage of water a distance, equivalent to about ten times the depth, down the estuary past a point. If the depth were 10 m and the current 100 cm sec^{-1}, the averaging time would need to be 100 seconds. It also appears that we really need to average over a longer time when the tidal current is low than when it is fast and also we need to measure over a longer period with increasing height above the bottom.

Experience, however, has shown that an averaging time of about $1-1\frac{1}{2}$ minutes is generally reasonable. In most cases this averaging is mainly visual and may be open to considerable errors. Once this averaging is complete then the effect of longer period turbulence appears in the tidal terms.

Mean Flow

To obtain the value for the mean flow we need a large number of equally spaced observations over one or two complete tidal cycles. For adequate representation of the mean, observations every half an hour can suffice, though in some cases it is possible to get away with hourly readings.

The mean flow can be obtained from a set of observations by dividing the tidal cycle into 12 lunar hours. From the data, current velocity profiles are drawn and the velocities interpolated and tabulated for each lunar hour or half lunar hour, for given fractions of the total water depth. As well as enabing determination of the mean velocity over a tidal cycle for each fraction of the water depth, the total discharge over a tidal cycle and at any time during the tide can also be calculated. The same process can be used to obtain mean salinities and temperatures.

The value for the mean flow $\bar{u}$, or non-tidal drift, obtained in this way and averaged over the cross-section, need not be equivalent to the river discharge. Because of the tidal fluctuation in cross-sectional area, the volume transports for a unit mean velocity will vary throughout the tidal cycle. If $u = \bar{u} + U_o \sin \omega t$, where u does not vary with width or depth, and $A = \bar{A} + A_o(\cos \omega t + \theta)$ where $\bar{A}$ is the mean cross-sectional area over a tidal period,

$$\text{river discharge} = \frac{1}{T} \int_0^T Au \, dt$$

$$= \bar{A}\bar{u} + \frac{1}{2\pi} \int_0^{2\pi} A_o \cos(\omega t + \theta) \, . \, U_o \sin \omega t \, . \, dt \quad (3.15)$$

If the tidal wave in the estuary is a standing wave and slack water occurs at both low and high water then $\theta = 0$, the second term is zero and the river flow equals the mean velocity $\bar{u}$ (sometimes called the non-tidal drift velocity). If the tidal wave is purely progressive then $\theta = \pi/2$ and the second term becomes $-\frac{1}{2} A_o U_o$ and the river discharge $R = \bar{A}\bar{u} - \frac{1}{2}A_o U_o$. Consequently the observed mean flow or non-tidal drift $\bar{u}$ can exceed the movement due to the river discharge divided by the cross-section area (Pritchard, 1958). The additional part of the non-tidal drift is compensation for inward mass transport due to a partially progressive nature of the tidal wave. In the Columbia estuary slack water lags the maximum and minimum cross-sectional area by $1-1\frac{1}{2}$ hours even at low river stage. As a consequence 30% of the non-tidal drift is compensation for the inward mass transport (Hansen, 1965).

Vertical Velocities

Because one of the primary actions in an estuary is the vertical entrainment of salt water, it is necessary to know the vertical velocities. Though it is possible to measure instantaneous vertical velocities it is at present impracticable to obtain sufficient measurements over an area and in time to be able to calculate the mean vertical velocity. Instead, the mean vertical velocities must be derived from the horizontal velocities using the principles of continuity.

As velocity is a vector having magnitude and direction, it can be represented by three components acting on mutually perpendicular axes. The x direction is taken longitudinally positive downstream, the y axis is lateral, positive to the right, and the z axis is vertically downwards. The values of longitudinal mean velocity $\bar{u}$ and lateral mean velocity $\bar{v}$ can be obtained from measurements at stations on cross-sections of the estuary. Then (similarly to the derivation of equation 5.1)

$$\frac{\partial \bar{u}}{\partial x} + \frac{\partial \bar{v}}{\partial y} + \frac{\partial \bar{w}}{\partial z} = 0 \qquad (3.16)$$

This is the equation of volume continuity and simply formalizes the fact that what goes in must come out. Any slowing down of water in the longitudinal sense requires additional discharge laterally or vertically. Using the horizontal mean velocities at stations on two cross-sections, then equation (3.16) can be solved, provided one boundary condition is known, that $\bar{w}$ is known at one depth. If tidal rise and fall have been eliminated by averaging over a tidal cycle, then a valid boundary condition is that $\bar{w}$ is zero at the sea surface when z is zero. The distribution of $\bar{w}$ with depth can then be calculated by stepwise integration from the surface downwards.

The vertical velocity may not be zero near the bed of the estuary as it may be equal to the vertical component of the mean longitudinal current flowing on a sloping bottom, though it will probably tend towards small values there, The vertical velocities are several orders of magnitude smaller than the longitudinal velocities, generally having values of the order 10^{-3} cm sec^{-1}.

Often equation (3.14) is used in a reduced form to calculate the vertical velocities. If $\bar{u}_A$ is the cross-sectional tidal mean velocity and the breadth b varies with depth and along the estuary, then:

$$\frac{\partial b \bar{u}_A}{\partial x} + \frac{\partial b \bar{w}}{\partial z} = 0 \qquad (3.17)$$

The vertical velocities are the mean over an area, parallel to the water surface between the two cross-sections and the estuary sides.

Secondary Flows

Some care must be taken in interpreting the results of analyses for the vertical velocities. Whereas $\bar{u}$ is the mean longitudinal velocity at a point, or if suitably averaged, the cross-sectional mean, the mean vertical velocity $\bar{w}$ is straightaway the mean averaged over a large area. The real vertical velocities may undergo drastic variation within that area because of the presence of secondary flows.

In meandering rivers the outside of a meander bend is usually occupied by a scour hole. Superimposed on the longitudinal flow is a secondary flow which tends to go downwards into the scour hole and upwards in the shallower water on the inside of the bend (Leopold and Wolman, 1960). This leads to a rotary secondary flow which is clockwise on left-handed bends and anti-clockwise on right-handed bends, looking downstream. The sense of rotation changes in the straight sections between meanders. There both rotations may be present, but separated in midstream by a shallower longitudinal bar. These secondary flows may be present even in straight channels (Einstein and Shen, 1964). There is evidence that when a salinity stratification is present the secondary flow system tends to be the opposite of that occurring in rivers, with upward flow occurring in the deeper water where the saline bottom water is thickest and downwards flow in the shallow water where the saline layer may be absent. The secondary flows produced by the meander topography are of the same order of magnitude as the vertical velocities of entrainment. Where entrainment is active upward flow will result.

In regions dominated by tidal movement there can be marked differences in the strength of ebb and flood currents at different positions in the channel. Again this situation is associated with meanders in the channel and is caused by the flood currents taking a straighter course than the ebb currents which tend to follow the meanders. These variations produce horizontal secondary circulation systems, which in the broader parts of the estuary may separate the flows into 'ebb and flood channels'. The former contain predominant ebb currents and shallow and narrow towards the sea. The latter narrow and shallow towards the land and contain predominant flood currents. Often these channels occur in pairs, one ending at a steep slope on the edge of the other. As the tidal excursion is generally longer than the length of one of these channels the water flows up one channel and down the other, forming a circulating system. The channels alter their position in an apparently consistent manner and the banks between vary in extent and volume. These movements can cause fluctuations of as much as 5% in estuary volume.

Consequently, in measurements of current velocity in estuaries one must be careful in the initial positioning of survey stations and cautious about taking cross-sectional averages of currents.

CHAPTER 4

Some Examples of Estuaries

In this chapter a number of typical estuaries of the different types will be examined. These examples have been chosen to illustrate the properties of estuaries with different tidal ranges, river flows and topographies and to extend the brief descriptions of the characteristics given in chapter 2. The measurements described will be used later in the sections dealing with detailed analysis of the salt balances and dynamics.

HIGHLY STRATIFIED ESTUARIES

The Vellar Estuary, India

The Vellar estuary (11°29' N lat., 79°47' E long.) has been described by Dyer and Ramamoorthy (1969). The section of the estuary investigated was at the lowermost 6 km, which is straight, about 300 m wide, and averaging about 1·5 m deep (Figure 4.1). Just inside the mouth lagoons exist on either side behind a wide sandy beach, the breakpoint bar forming the essential feature of a bar-built estuary. Within the estuary there is a series of alternating scour holes that are consistent with a meandering flow of water. The normal tide range in the estuary is about 70 cm.

Three surveys were completed to examine the distribution of salinity, temperature and current velocity within the estuary. Only the lower 10

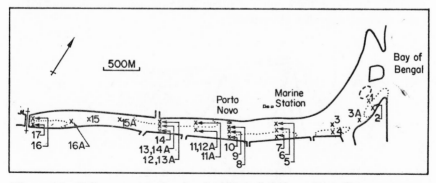

Figure 4.1 Vellar estuary, showing location of stations. Dotted areas represent the approximate limit of deep-water areas. (Reproduced with permission from K. R. Dyer and K. Ramamoorthy, 1969, *Limnol. Oceanog.*, **14**, 4–15, Figure 1)

33

stations were occupied on 20th January, 1967, which was a week after heavy rain had caused the river to flood. On 27th–28th January and 9th–10th February, surveys were completed over the entire 17 stations. On 15th February a detailed survey to analyse the mixing process was carried out with 15 measurements at each of three stations on two sections of the lower estuary at hourly intervals.

Distribution of salinity, temperature and currents

The distribution of salinity at high and low waters for the three periods is shown in Figures 4.2, 4.3 and 4.4. During the flood tide there was an intrusion

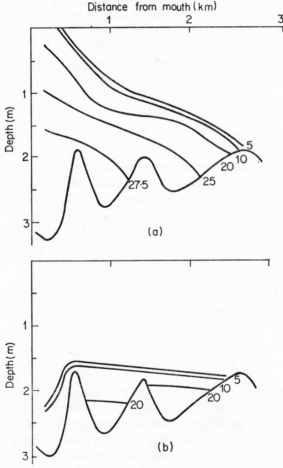

Figure 4.2 Longitudinal distribution of salinity on 20th January, 1967. (a) High water. (b) Low water. Contours in ‰. (Reproduced with permission from K. R. Dyer and K. Ramamoorthy, 1969, *Limnol. Oceanog.*, **14**, 4–15, Figure 2)

of a saline wedge along the bottom, bounded by a halocline of up to 20‰ in 0·5 m, from the fresh river water on the surface. This wedge acted as a piston impounding river water in the upper estuary. At high water at the mouth, the water column was only moderately stratified with a bottom salinity that was increasing during the surveying period. Only a short distance inside the mouth fresh water was present on the surface though the distance of penetration of

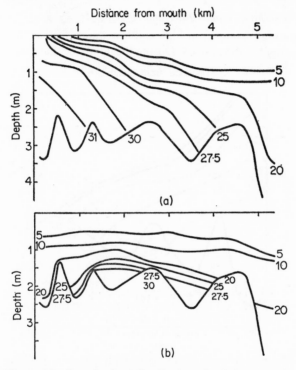

Figure 4.3 Longitudinal distribution of salinity on 27th–28th January, 1967. (a) High water. (b) Low water. Contours in ‰. (Reproduced with permission from K. R. Dyer and K. Ramamoorthy, 1969, *Limnol. Oceanog.*, **14**, 4–15, Figure 3)

the salt wedge increased with time as the river discharge decreased from the initial flood stage. At about high water the impounded river water was released and quickly established a homogeneous surface layer with a virtually horizontal halocline throughout the estuary. This halocline was rapidly pushed downwards until it reached the level of the sills between the basins, the shallowest sill being 0·5 km inside the mouth (Station 3A). The saline waters were then isolated within the basins, but under the influence of the fresh water flow the saline water was entrained from the top of the halocline and flowed over the sills from one basin to another.

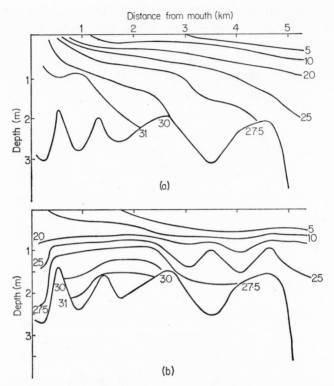

Figure 4.4   Longitudinal distribution of salinity on 9th–10th
February, 1967. (a) High water. (b) Low water. Contours in ‰.
(Reproduced with permission from K. R. Dyer and K. Rama-
moorthy, 1969, *Limnol. Oceanog.*, **14**, 4–15, Figure 4)

Certain of the basins seemed to be especially prone to flushing out by the
fresh water. It is apparent by comparison between the high and low tide
bottom salinities, that the most stable basins were those by the Biological
Station (Stations 5, 6, 7) and by the railway bridge (Stations 16, 17). The
basins most prone to dilution of bottom salinity were those at Stations 2 and
4. These effects may have been caused by peculiarities in the water movement
as a result of the bottom configuration.

Just before low water, as the downstream pressure gradient diminished,
the seaward flowing surface water decreased in velocity and the saline water
started moving landwards along the bottom at the mouth. It then passed
upstream from one basin to the next, causing abrupt increases in bottom
salinity. Development of this salt wedge pushed the halocline upwards and
impounded the river flow in the upper estuary.

These processes are further exemplified by consideration of the variation
in bottom salinity at various points in the estuary. The tidal variation of

bottom salinity on 9th February is shown in Figure 4.5. The time between consecutive high waters was divided into 12 lunar hours. At Station 2 there was a gradual reduction of bottom salinity from just after high water to just before low water, when the inflow of saline bottom water began. At Station 3A the bottom salinity was slightly less than that at Station 2 except

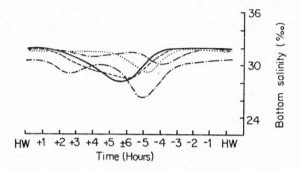

Figure 4.5 Variation in bottom salinity on 9th February, 1967. —— Station 2; - - - Station 3A; ... Station 4; ----- Station 7; ---- Station 9. (Reproduced with permission from K. R. Dyer and K. Ramamoorthy, 1969, *Limnol. Oceanog.*, **14**, 4–15, Figure 5)

between HW + 4·5 and HW − 5·75. This was the period during which the saline water flowing over the sill was supplying the bottom water at Station 2. At Station 4 the bottom salinity was almost constant between HW − 3 to HW + 5, when there was a rapid reduction in bottom salinity owing to increased flushing by the fresher surface flow. Then at HW − 4·75 the inflowing saline water had penetrated up from Station 3A, leading to a rapid increase in bottom salinity. At Station 7 the pattern was similar except that there was a further 0·75 hour delay in the inflowing salt water. Also there was a slight minimum bottom salinity at HW + 4 when the outflowing ebb current penetrated right to the bottom at its maximum amplitude. As soon as the pressure of the current eased, the isohalines became less steeply inclined, the bottom water creeping further up the landward edge of the basin and slightly enhancing the bottom salinity. Station 9 was situated on the sill between two basins and reacted differently as a result of the salt water flowing out of the basin above under increasing current velocity, starting at HW + 2·5. After HW + 4, as the current diminished, this flow decreased and the salinity fell until the current reversed an hour after low water.

The relationship between salinity and velocity variations during a tidal cycle is shown by the results for Stations 2 and 7 on 17th January (Figure 4.6). At Station 2 on the flood tide, the current was maximum at mid-depth and

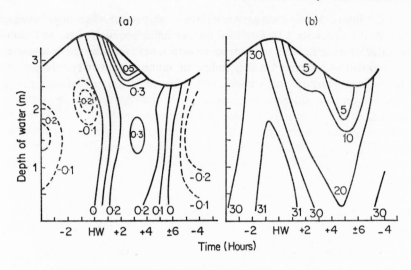

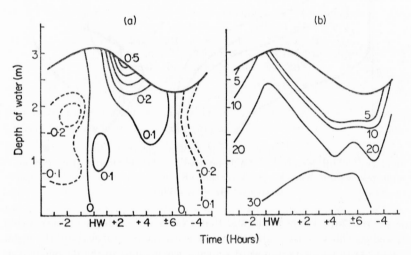

Figure 4.6   Variation of: (a) longitudinal velocity (m/sec); (b) salinity (‰).
Top, Station 2 on 27th January, 1967; bottom, Station 7 on 27th January,
1967. (Reproduced with permission from K. R. Dyer and K. Ramamoorthy,
1969, *Limnol. Oceanog.*, **14**, 4–15, Figures 6 and 7)

salinity increased regularly from top to bottom. About an hour before high
water, the surface current reversed and the salinity started decreasing. The
bottom current, however, reversed at high water. During the ebb tide, the
maximum velocity was at the surface, but with measurable velocities at the
bottom. About 0·75 hour before low water, the saline water started flowing
in along the bottom, but the surface current continued to flow seaward until

about 0·5 hour after low water. At Station 7 the pattern was similar except that the current reversed throughout the column before high water and after low water. During the ebb, the bottom velocities were immeasurably low and this coincided with the period of high bottom salinities caused by entrapment of the bottom water.

The temperature distribution appeared to be dominated by diurnal variation. On 15th February, for instance, high water occurred at 12.15 hours and the maximum temperature of 28·7°C was recorded on the surface at about 15.30 hours. The minimum 25·5°C occurred at mid-depth at 07.30 hours. The more saline bottom water, however, was generally slightly colder than the fresher surface water.

Sufficient measurements were taken to enable calculation of the mean salinity and velocity over a tidal cycle at each station for the different periods. Typical curves are shown for Station 7 (Figure 4.7). The mean salinity at the

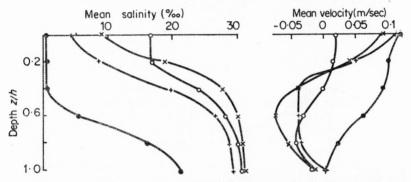

Figure 4.7 Tidal mean salinity and longitudinal velocity for Station 7. ● 20th January; +27th January; ×9th February; ○ 15th February. (Reproduced with permission from K. R. Dyer and K. Ramamoorthy, 1969, *Limnol. Oceanog.*, **14**, 4–15, Figure 9)

first period (20th January) showed that a fresh surface layer overlay a thin salt wedge. This was associated with a seaward velocity throughout the water column. At this time the seaward velocity was greatest in the deeper part at Station 7. In the shallower parts of the section the seaward mean velocities were less. Thus the isotachs were tilted downwards towards the right. As river discharge decreased, the saline bottom layer became thicker and more homogeneous and a landward mean velocity developed at mid-depth. At the last period the surface mean velocity had decreased abruptly and at Station 7 was considerably less than the landward bottom flow. The necessary mean seaward flow due to river discharge was maintained by increased mean surface flow and decreased mean landward bottom flow in the shallower waters of Stations 5 and 6. Thus with decreased river discharge the depth of no motion and the isotachs became tilted downwards towards the left.

The distribution of depth mean salinity for the various periods is shown in Figure 4.8. The salinity gradient decreases with time as the river discharge decreases from the flood stage. The effect of the entrapment of bottom water is shown in the higher mean salinities of the deeper water areas. It is possible that even at flood stages saline water may exist in the deepest area by the railway bridge (Station 17).

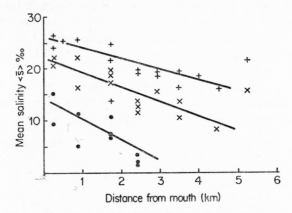

Figure 4.8 Longitudinal distribution of tidal-depth–mean salinity. ● 20th January; × 27th–28th January; + 9th–10th February. (Reproduced with permission from K. R. Dyer and K. Ramamoorthy, 1969, *Limnol. Oceanog.*, **14**, 4–15, Figure 10)

The three lines have the formulae:

$$\langle \bar{s} \rangle = 15{\cdot}6 - 5{\cdot}4x \quad \text{for river flow of } 375 \text{ m}^3 \text{ sec}^{-1}$$
$$\langle \bar{s} \rangle = 22{\cdot}4 - 3{\cdot}1x \quad \text{for river flow of } 90 \text{ m}^3 \text{ sec}^{-1}$$
$$\langle \bar{s} \rangle = 26{\cdot}5 - 2{\cdot}2x \quad \text{for river flow of } 86 \text{ m}^3 \text{ sec}^{-1}$$

where x is distance in kilometres from the mouth. These three lines intersect about 6·5 km seaward of the mouth, indicating that estuarine mixing processes can extend well into the open sea.

Water circulation

The mean vertical velocities were calculated for 0·5 m intervals for Stations 5–10, from the equation of continuity (Figure 4.9). For the high river discharge on 20th January, the vertical velocities indicated the presence of a downward flow at Stations 7 and 10 and an upward flow in the shallower water at Stations 5 and 8. Associated with these, a right lateral mean velocity at the surface and a left lateral mean velocity at the bottom were measured. This clockwise secondary current which was, of course, superimposed on the mean seaward flow, is typical of that associated with the scour holes on the meanders of alluvial channels. As the river flow decreased, the saline intrusion developed. For the 15th February the vertical velocities show that there was an upward flow in the deeper areas and a downward flow in the shallower areas. These were associated with a measured left lateral mean flow at the surface and a right lateral mean flow at the bottom. Thus, as the saline intrusion developed, a counterclockwise secondary current system formed. Between these two basic flow patterns, the measurements of 9th February showed

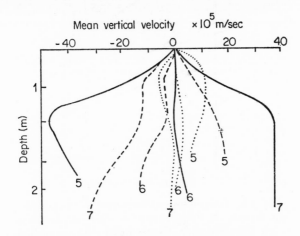

Figure 4.9 Calculated mean vertical velocities for Stations 5–7. —— 20th January; - - - 9th February; ... 15th February 1967. (Reproduced with permission from K. R. Dyer and K. Ramamoorthy, 1969, *Limnol. Oceanog.*, **14**, 4–15, Figure 11)

a more complicated circulation. At that time the counterclockwise saline wedge pattern was well developed at the lower section (Stations 5–7), but at the section of Stations 8–10, where the saline underflow was not fully developed, the pattern was less well established.

Classification of the Vellar Estuary

The values of the dimensionless stratification and circulation parameters were obtained for the section of Stations 5–7 from the observed salinity and velocity values. The results are plotted in Figure 4.10 with the calculated river discharge rates in $m^3 \ sec^{-1}$ in parentheses. This shows that at periods of high river discharge the estuary is a salt-wedge type. At decreased river flow the estuary becomes well stratified with the mean current reversing at depth. It is probable that as river flow diminishes further the estuary would become less well stratified. In the hot season, with no river flow, the estuary may become homogeneous, though temperature may become important in producing density differences by surface heating.

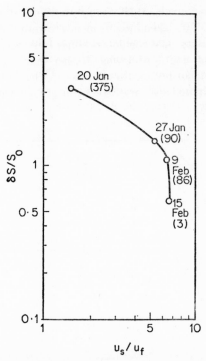

Figure 4.10   Classification of Vellar estuary according to method of
Hansen and Rattray (1966). River discharge rates (m³/sec) in paren-
theses. (Reproduced with permission from K. R. Dyer and K.
Ramamoorthy, 1969, *Limnol. Oceanog.*, **14**, 4–15, Figure 12)

Columbia River, U.S.A.

The Columbia River discharges into the Pacific Ocean at 46°15′ N and
124° W and is described by Hansen (1965). The tides at the mouth have a
range between 2·3 m and 4 m and propagate nearly 300 km upstream during
low river flows. The tides reverse the river current in the lowest 40–80 km.
The mean annual discharge is about 7300 m³ sec^{-1} but ranges between
3000 and 20000 m³ sec^{-1}. The normal maximum occurs in mid-May to mid-
June, resulting from the interior snow melt. Coastal rain can give maxima
during the winter. Although the tides are moderately strong, the flow ratio
(ratio of river discharge during a tidal cycle to the tidal prism) is between 1
and 0·2, making the river current 0·5–0·1 of the tidal currents at the mouth.
Even at low river discharge the discharge is strong enough to cause high
slack water and low slack water to lead and lag respectively the tidal amplitude
by about 25 minutes.

A measurable salinity intrusion only extends about 10 km into the estuary
at high river discharge and rarely greater than 20 km for any river flow:

much the same order as the tidal excursion. Surface to bottom salinity differences can exceed 20‰. There is a tendency toward the salt-wedge type of behaviour, but the strong tidal currents produce large vertical mixing. At the mouth the right-hand side is somewhat fresher than the left at high river discharge. This situation is reversed at low run-off. The salinity can range from zero at low water with high discharge to oceanic at high water and low run-off (Figure 4.11).

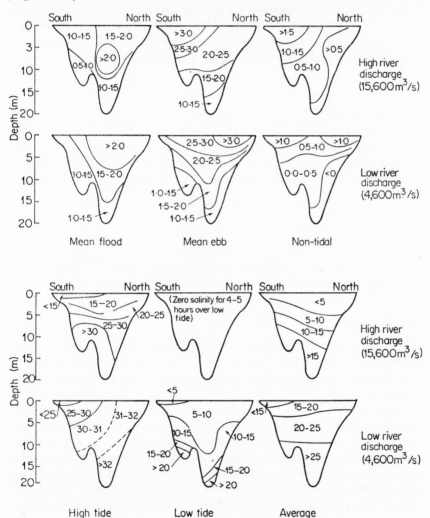

Figure 4.11 Velocities and salinities on a cross-section at the mouth of the Columbia River. Top, current speed in knots (positive non-tidal current indicates seaward flow). Bottom, salinity in ‰. (Reproduced with permission from D. V. Hansen, 1965, *Ocean Science and Ocean Engineering*, 943–955, Figures 2 and 3)

The ebb current frequently exceeds five knots at the surface and upstream net flow near the bottom is only found at the mouth at times of low river discharge (Figure 4.11). Reversal of net flow does not occur at times of high river discharge, but the zone of low mean velocity along the right-hand side appears to be a continuing manifestation of the density gradient. The appearance of the inward flow on the right-hand side, contrary to that expected under the influence of Coriolis force, must be caused by topography.

The Columbia River estuary is characterized by an unusual combination of large river discharge and strong tidal currents. Upstream salt transport must be effected mainly by diffusion. This estuary is a borderline case between a salt wedge estuary and a highly stratified estuary with landward net flow.

FJORDS

Silver Bay, Alaska

Measurements in Silver Bay have been reported by McAlister *et al.* (1959). The fjord is U-shaped in cross-section, about five miles long and half to three-quarters of a mile wide. There is no entrance sill, but a small sill at a depth of about 70 m occurs a mile inside the mouth. The main basin has a depth of

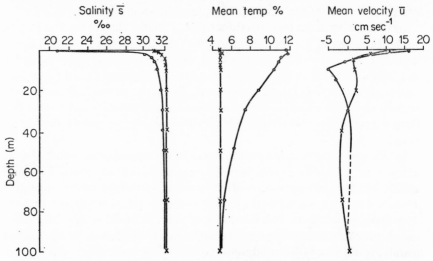

Figure 4.12   Mean salinity, temperature and velocity, Silver Bay, Alaska, centre section. ○ High run-off, July, 1956, × low run-off, March, 1957. (Reproduced with permission from W. B. McAlister, M. Rattray Jr. and C. A. Barnes, 1959, *Tech. Rept.* 62, University of Washington, Oceanography Dept., Figures 9 and 10)

about 80 m. The Bay opens onto the Eastern Channel of Sitka Sound which is 120–140 m deep. The mean tidal range is about 2 m. High run-off reaching about 86 m^3 sec^{-1} occurs in late spring and summer, with low run-off, less than 3 m^3 sec^{-1}, occurring in the winter.

Measurements were mainly completed on three sections across the mouth over a number of tidal cycles at both high and low river discharge. Salinity and temperature measurements were completed at stations on all three sections, but current measurements were only completed at the centre section (section 2). Values for mean salinity, temperature and current velocity at the centre section are shown in Figure 4.12.

During high discharge conditions the outflowing surface water layer was about 5 m thick and its salinity ranged from zero at head to 28‰ at the mouth. A compensating landward flow occurred between 5–30 m depth. During the low discharge stage, conditions were more homogeneous. The outflowing layer occupied the surface 30 m and the inflow was between 30 m depth and the bottom. The amplitude of the tidal velocity was about 2 cm sec^{-1}, considerably smaller than the mean flow in the surface layer.

Hardanger Fjord, Norway

The hydrography of Hardanger Fjord has been described by Saelen (1967). This fjord is over 100 km long and over 850 m deep and has a sill depth of about 150 m. River discharge averages 385 m^3 sec^{-1} during June, but is less than 20 m^3 sec^{-1} during February. The longitudinal distribution of temperature and salinity at high and low flows are shown in Figure 4.13.

In summer, with the high river discharge, a fresher surface layer is formed with a thickness of less than 20 m. At this time the surface water is warmer than that below. There is a possibility of an inflow of coastal water occurring at a depth of 10–100 m during the late summer, providing an additional heat supply to the surface layers. During the winter the surface salinity is only slightly lower than that at depth. The surface temperature is also low and there is a possibility that vertical thermohaline mixing may occur if the surface density is high.

In the winter bottom temperatures are lower towards the mouth. In the summer the temperatures near the bottom at the mouth are below the winter values, but little change has occurred in the temperatures further inland. It appears that there is a summer inflow of colder water, at sill depths, into the deep water of the fjord. This renewal may occur annually or less frequently, and is probably independent of the upper-layer circulation, being affected mainly by offshore water conditions.

Alberni Inlet, British Columbia

Tully (1949) has described extensive measurements in Alberni Inlet. The low salinity surface layer has a depth of about 5 m and is fairly homogeneous near the head of the inlet. When discharge was less than 74 m^3 sec^{-1} the surface layer became shallower and fresher, with decreasing discharge. Above this discharge, the upper zone becomes deeper and more saline with increasing flow. The amount of fresh water contained in the upper zone was

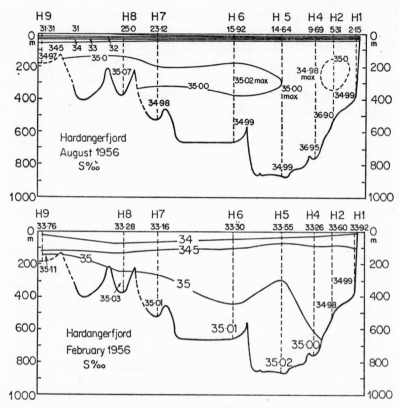

Figure 4.13A Distribution of salinity in a longitudinal section of Hardanger Fjord. Top, August, 1956; bottom, February, 1956. (Reproduced with permission from O. H. Saelen, *Estuaries* (ed. G. H. Lauff), AAAS Pub. No. 83, 1967, Figures 2 and 3. Copyright 1967 by the American Association for the Advancement of Science)

about 90% of the total at any section, and was a constant proportion in spite of differences in tidal amplitude, river discharge and position.

Near the inlet head the entry of the river flow caused complicated flow patterns, but lateral homogeneity was attained within a few miles. During the tidal cycle the lower boundary of the fresh surface layer was alternately tilted toward the head and the mouth at high and low tide. In the lower reaches, however, the boundary remained horizontal as the tidal deformation could be relieved by local accelerations.

Knight Inlet, British Columbia

Pickard and Rodgers (1959) have described extensive current measurements at two stations in Knight Inlet. Station $3\frac{1}{2}$ was situated on a sill with a depth of about 75 m and Station 5 was in a water depth of about 350 m in the inner

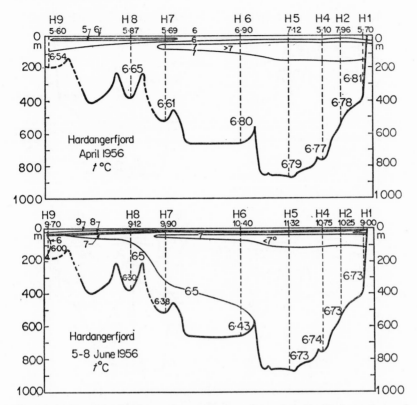

Figure 4.13B Distribution of temperature in a longitudinal section of Hardanger Fjord. Top, April 25th, 1956; bottom, June 6th, 1956. (Reproduced with permission from O. H. Saelen, *Estuaries* (ed. G. H. Lauff), AAAS Pub. No. 83, 1967, Figures 13 and 14. Copyright 1967 by the American Association for the Advancement of Science)

basin of the Inlet. The maximum tidal range is about 5 m. Currents at all depths at Station $3\frac{1}{2}$ showed tidal oscillations with peaks midway between predicted high and low water, i.e. a standing wave oscillation, but there were large irregularities in the currents at all depths. The amplitude of the tidal currents was about 50 cm sec^{-1} and changed little with depth. During the first 25 hours of measurement when the wind was light, there was a net outflow down to 40 m and a landward flow beneath. In the final 25-hour period with a 6 m sec^{-1} up-inlet wind, the surface flow was landward to a depth of 6 m and seaward between 6 m and 55 m (Figure 4.14a).

At Station 5 the deeper current measurements showed tidal currents up to about 15 cm sec^{-1}. Slack water coincided with predicted high and low water, but there were again large irregularities in the flow. In the surface layer, however, the currents were several times larger, ranging between 120 cm sec^{-1}

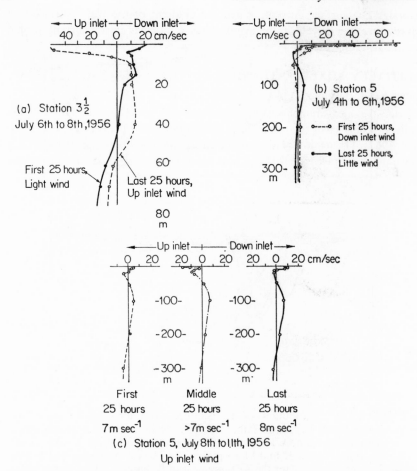

Figure 4.14 Current measurements with varying wind conditions, Knight Inlet. (Reproduced with permission from G. L. Pickard and K. Rodgers, 1959, *Journal Fish. Res. Bd. Canada*, **16**, 635–678, Figures 13, 16 and 20)

down-inlet to 45 cm sec⁻¹ up-inlet. Strong surges in the current occurred in the hour before high water. At intermediate depths (10–50 m) currents were irregular, sporadic and often zero. The flood current at 50–100 m started after predicted high water and extended down to 300 m at predicted low water. For the first 25 hours there was a down-inlet wind of 5 m sec⁻¹, but during the last 25 hours there was little wind. The profiles of mean current (Figure 4.14b) show a faster, deeper surface flow with the down-inlet wind and a change in the mean flow at depth.

Further measurements at Station 5 gave similar results, but in this case there was a continuous up-inlet wind which increased to 12 m sec⁻¹ at one stage. The mean current profiles (Figure 4.14c) show the increased wind speed reversing the surface current. When the gradient of the water surface was adjusted to the effect of the wind stress a normal mean current profile became re-established.

PARTIALLY MIXED ESTUARIES
James River, Virginia

Extensive measurements were undertaken in 1950 on the James River Estuary. Analyses of the results have been reported by Pritchard (1952, 1954, 1956, 1967). The section of the estuary investigated was between 20 and 45 km above the mouth (Figure 4.15). A number of stations were occupied for three

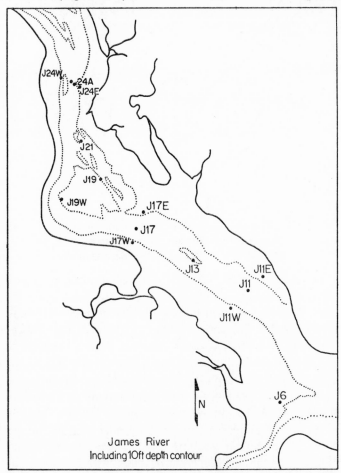

Figure 4.15 James River, topography and Station positions. (Reproduced with permission from D. W. Pritchard and R. E. Kent, 1953, *Tech. Rept.* VI, Chesapeake Bay Inst., Johns Hopkins University, Figure 1)

periods of at least four days during which serial measurements of current, salinity and temperature were taken. Several stations were concentrated on three cross-sectional traverses. The velocities were obtained using a current drag (Pritchard and Burt, 1951). The surface high and low water salinities for one day are shown in Figure 4.16. The surface salinities are noticeably

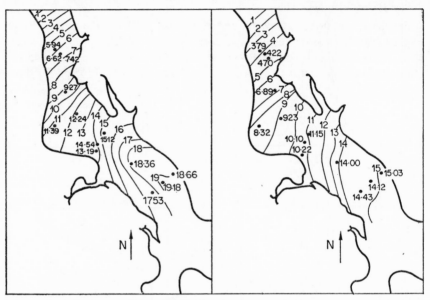

Figure 4.16. Surface salinity distribution in the James River estuary: (left) high water: (right) low water. (Reproduced with permission from D. W. Pritchard, 1952a, *Journal Marine Res.*, **11**, 106–123, Figure 4)

lower on the right-hand side. At other times similar distributions were observed with surface salinities at high water being about 4‰ higher than those at low water. The mean vertical salinity profiles at Station J17 show a mid-depth halocline and a difference of about 2‰ between the surface and the lower layers (Figure 4.17). The mean velocities show a two-layer flow with an upsteam flow in the bottom layer and a downstream flow in the surface layer (Figure 4.17). The mean of the seaward velocity in the surface layer was about 19% of the average tidal velocities, whereas in the lower layer the landward flow was about 20% of the average tidal velocities. The mean flow summed over depth was much larger than that due to the seaward movement of river flow. At Station J17 the level of no net motion sloped by over 1 m downwards towards the right-hand side of the estuary.

Using the differences in mean horizontal velocity between the three cross-sections, mean vertical velocities were calculated. The vertical velocities calculated at Station J17 are upward at all depths with a maximum value of about 1×10^{-5} m sec^{-1} at mid-depth (Figure 4.17).

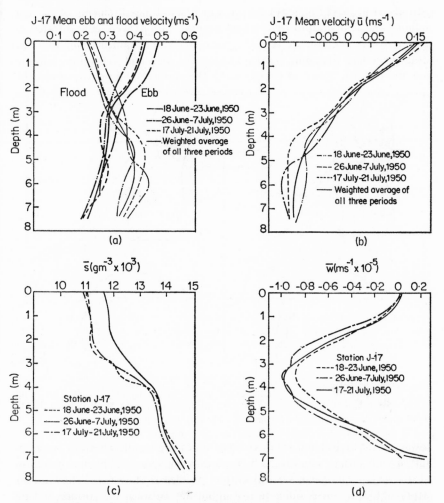

Figure 4.17 James River, Station J–17. (a) Vertical profile of mean ebb and flood currents. (b) Vertical profile of net non-tidal velocity. Net flow is seaward in the upper layer (positive values) and up-estuary in the lower layer (negative values). (c) Vertical profile of mean salinity. (d) Mean vertical velocity ($\overline{w}$). (Reproduced with permission from D. W. Pritchard and R. E. Kent, 1953, *Tech. Rept.* VI, Chesapeake Bay Inst., Johns Hopkins University, Figures 2, 3, 4, 6)

This set of observations is probably the most extensive and the most thoroughly analysed from any estuary. As we shall see later, many other investigations have been greatly influenced by this study.

Mersey Narrows, England

The Mersey Narrows is a section of the river Mersey which is about 10 km

long and 1 km wide and which connects Liverpool Bay with a wider and
shallower inner estuary (Figure 4.18). The maximum depth in the Narrows
exceeds 20 m. The tide range can reach 9·4 m and produces currents up to
2·2 m sec^{-1}. The river flow is small compared with the large tidal flow, but is

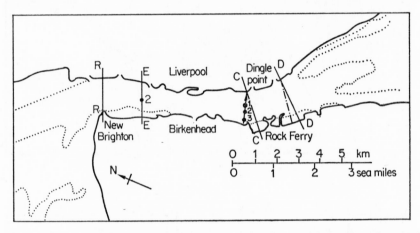

Figure 4.18 Narrows of the Mersey estuary, showing positions of stations and
sections. . . . , Low water line, spring tides. (Reproduced with permission from
K. F. Bowden and S. H. Sharaf el Din, 1966, *Geophy. Jour.*, **10**, 383–399, Figure 1)

very variable ranging from about 25 m^3 sec^{-1} to over 200 m^3 sec^{-1}. The
estuary has been described by Hughes (1958) and Bowden and Sharaf el Din
(1966a).

The longitudinal variations of salinity at high and low water are shown in
Figure 4.19. The difference between extreme values of salinity at any position,
during a tidal cycle, was about 4‰ for a low river discharge period. For high
river discharge the difference was about 11‰. During low river discharge the
salinity differences between surface and bottom were not a maximum at high
or low water, but continued to increase to about 1‰ about 40 minutes after
high water and to more than 1‰ over an hour after low water. The variation
of salinity with depth was small or absent when the tidal current was strong,
so that for the mid-tide period the isohalines were almost vertical. For high
river discharge a vertical salinity gradient was always apparent and there
was no period when the isohalines were vertical, also the differences in
surface to bottom salinities at slack water were larger than for low river
discharge.

The distribution of depth–mean salinity for two periods is shown in
Figure 4.20. For low river discharge the formula is $\langle \bar{s} \rangle = 30\cdot11 - 0\cdot236x$ and
for high river discharge $\langle \bar{s} \rangle = 27\cdot36 - 0\cdot375x$.

Low salinity surface waters are likely to exist well into Liverpool Bay and

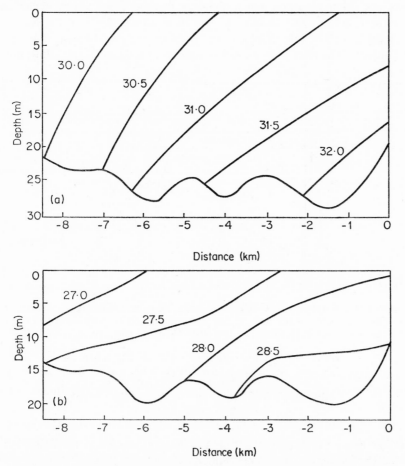

Figure 4.19 Isohalines (a) at high water, (b) at low water, in a vertical section along the middle of the Narrows (distance in km from Rock Light, New Brighton). (Reproduced with permission from P. Hughes, 1958, *Geophy. Jour.*, **1**, 271–283, Figures 3 and 4)

offshore observations of currents and salinity have shown that an estuarine-type circulation extends to a distance of at least 12 miles from the mouth of the Mersey (Bowden and Sharaf el Din, 1966b).

Current and salinity measurements at stations on section C have been reported by Bowden and Sharaf el Din (1966a) and at stations along the Narrows by Bowden (1963). The latter measurements showed a surface outflow and a bottom inflow, but with variations in the depth-mean of the residual flow between $1 \cdot 0$ to $-8 \cdot 7$ cm sec^{-1}. These values are too large to represent the mean flow through a complete section and were probably caused by lateral variations. At each station there was a downstream flow near the

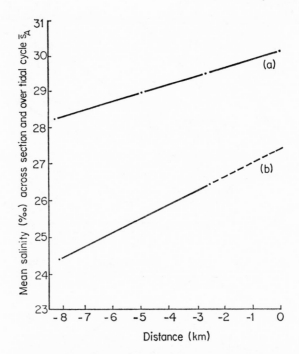

Figure 4.20 Distribution of mean salinity along the Narrows (distance in km from Rock Light, New Brighton) (a) 1956 May, (b) 1957 March. (Reproduced with permission from P. Hughes, 1958, *Geophy. Jour.*, **1**, 271–283, Figure 6)

surface and an upstream flow near the bottom, relative to the depth mean flow. The former measurements showed a depth mean flow of 7·7 cm sec^{-1} upstream, 0·9 cm sec^{-1} downstream and 10·4 cm sec^{-1} downstream at stations 1–3 respectively (Figure 4.21). The lateral mean velocities indicated a considerable eastward flow at all stations which may be associated with the curvature of the channel near this section. There were indications, however, of an anticlockwise secondary circulation westwards on the surface and eastwards near the bottom. The mean salinity profiles for this period (Figure 4.21) showed that the fresher water was concentrated on the left-hand side of the section.

Salinity observations over the period October, 1962 to October, 1963 were completed at a number of stations within the Narrows (Bowden and Sharaf el Din, 1966a). It was found that the best correlation between the salinities and the river flow occurred using the average river flow for the week ending on the day of salinity sampling. This is obviously a measure of the residence time of the fresh water within the estuary.

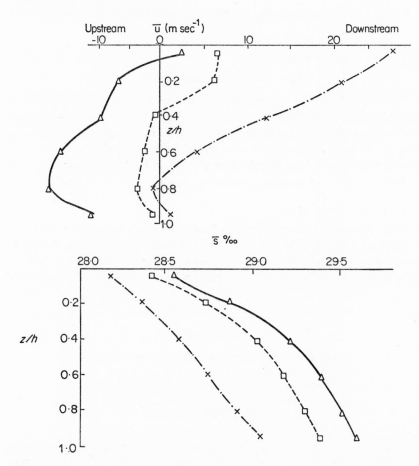

Figure 4.21 Mean velocity and salinity at stations on section C, Mersey Narrows. △ Station 1, □ Station 2, × Station 3. (Reproduced with permission from K. F. Bowden and S. H. Sharaf el Din, 1966, *Geophy. Jour.*, **10**, 383–399, Figures 2 and 4)

Assuming a linear relationship between salinity (‰), river discharge R (m³ sec⁻¹) and tidal range T (ft).

Mean salinity at the centre of section EE

$$S_{EE} = 30 \cdot 79 - 0 \cdot 0288R - 0 \cdot 0555T$$

Mean salinity at the centre of section DD

$$S_{DD} = 29 \cdot 45 - 0 \cdot 0411R - 0 \cdot 0393T$$

Horizontal difference of salinity between the centres of sections EE and DD

$$\Delta S = 1 \cdot 27 + 0 \cdot 0133R - 0 \cdot 0156T$$

Vertical difference of salinity, surface to bottom, at the centre of section EE

$$\delta S = 0{\cdot}36 + 0{\cdot}019R - 0{\cdot}0125T$$

and at the centre of section DD,

$$\delta S = 0{\cdot}47 + 0{\cdot}0264R - 0{\cdot}0222T$$

This shows that the horizontal and vertical gradients of salinity depend on the river flow more than on tidal range. However, Bowden and Gilligan (1971) found that the density current transport was relatively insensitive to changes in river discharge, but increased with increasing tidal currents. Similar results have been found for suspended sediment transport (Price and Kendrick, 1963), the transport increasing rapidly with tidal amplitude.

Classification of the Mersey

From extensive measurements on four sections in the Mersey Narrows, Bowden and Gilligan (1971) have calculated $\delta S/S_0$ and u_s/u_f and plotted the results on a stratification–circulation diagram (Figure 4.22). The results for each section show considerable scatter, possibly due to non-steady state

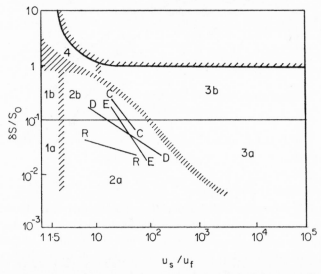

Figure 4.22   Stratification–circulation   diagram   for   the Mersey. Lines CC, DD, EE and RR are taken from Figure 4.18. (Reproduced with permission from K. F. Bowden and F. M. Gilligan, 1971, *Limnol. Oceanog.*, **16**, 490–502, Figure 7)

conditions operating, and also lateral variations in excess of those allowed for in the calculations. However, lines through the points lie almost parallel to each other, yet are displaced from each other. All of the lines lie within the Type 2 classification in which the net flow reverses with depth and both diffusion and advection are important in the upstream flux of salt. The farthest seaward section lies entirely within the 2a category where the stratification is relatively slight. The other sections extend into the 2b region where the stratification is greater.

Southampton Water, England

The estuary of Southampton Water is the drowned lower portions of the Test and Itchen Rivers, forming a typical coastal plain estuary (Figure 4.23). The maximum combined river flow of about $28 \cdot 3$ m^3 sec^{-1} occurs in the late winter and gradually diminishes to about 14 m^3 sec^{-1} through spring and

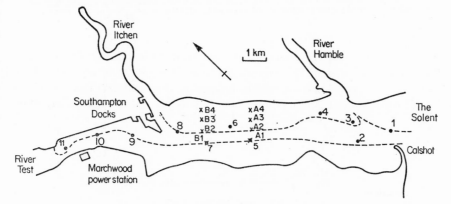

Figure 4.23 Southampton Water showing station positions: - - -, limits of deep water channel, 6 m depth

summer. The Test has about twice the flow of the Itchen. The tidal range within the estuary varies from about 5 m to about $1\frac{1}{2}$ m.

Salinity and current distribution

Figure 4.24 shows the high water surface salinity distribution on 25th–26th May, 1966, together with a longitudinal section where the salinity is projected onto the centre line of the estuary without adjustment of depth. Below Station 7 the water column was almost homogeneous with salinity increasing only gradually towards the sea. There was, however, quite a considerable gradient across the estuary with the fresher water on the left-hand side. This is opposite to that expected under the influence of Coriolis force and may be related to topographic effects. Above Station 7 the stratification increased towards the head of the estuary, but a salinity minimum occurred in the upper part of the water column between Stations 8–11. This was particularly marked at Station 9 where there was a salinity difference of about 3‰ between the surface and the bottom. This minimum was associated with the outfall of Marchwood Power Station situated on the right bank of the estuary. The

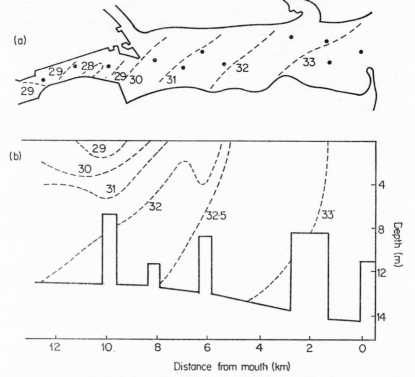

Figure 4.24 Southampton Water. Salinity distribution at high water, 25th–26th
May, 1966. (a) Surface salinity ‰, (b) longitudinal section.

cooling water intake is above Station 11 and fresher water taken in there is released between Stations 10 and 11, resulting in a fresh water depleted area between intake and outfall. The outfall water must mix down and across the estuary and be completely assimilated by about Station 7.

The surface and longitudinal salinity distribution for low water on the 25th–26th May, 1966 is shown in Figure 4.25. The lower part of the estuary maintained a low longitudinal salinity gradient, but the water column was stratified throughout. There was still concentration of fresher water on the left-hand side. Above Station 7 there was an abrupt decrease in the near surface salinity to the area of the power station outfall and again an increase in salinity between the intake and outfall. The maximum vertical salinity gradient had increased to over 10‰ at Station 10, caused mainly by the reduction in surface salinity. The salinities near the bottom changed little from high to low water.

The patterns of salinity distribution for other periods of measurement are similar, showing that at high water the lower part of the estuary is fairly homogeneous vertically, though stratification increases towards the head of

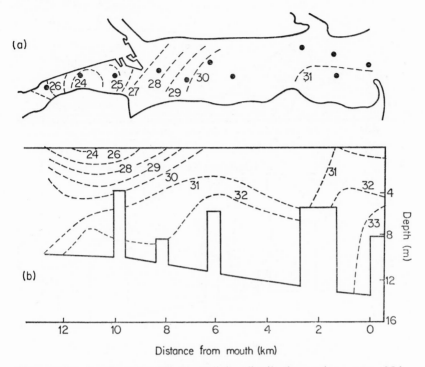

Figure 4.25 Southampton Water. Salinity distribution at low water, 25th–26th May, 1966. (a) Surface salinity ‰, (b) longitudinal section

the estuary. At low water the estuary is stratified throughout. The lateral salinity gradient increased with increased river discharge and with decreasing tidal range. The maximum lateral gradient at Station 7 was about three times the longitudinal salinity gradient.

The velocity variation during the tidal cycle is complex and is related to the unusual tide curve (Figure 4.26). There are two high waters about 2 hours apart, with a drop of about 15 cm between them at spring tides. A fast 4-hour ebb follows until low water, with a maximum velocity of about 1 m sec^{-1} on spring tides. The flood tide flows in two periods from low water to about $3\frac{1}{2}$ hours before high water and from about $2\frac{1}{2}$ hours before high water until high water, separated by a 'young flood stand'.

The water movements at high water are also complex, depending on the difference in height of the double high waters and the magnitude of the recession between. Generally, however, in the upper estuary the bottom current reverses about 50 minutes after high water, whereas in the lower estuary the bottom water floods until about $1\frac{1}{4}$ hours after high water. (This did not happen on 9th August 1966). The saline bottom water starts flowing landwards up to 20 minutes before low water at the mouth, whereas the

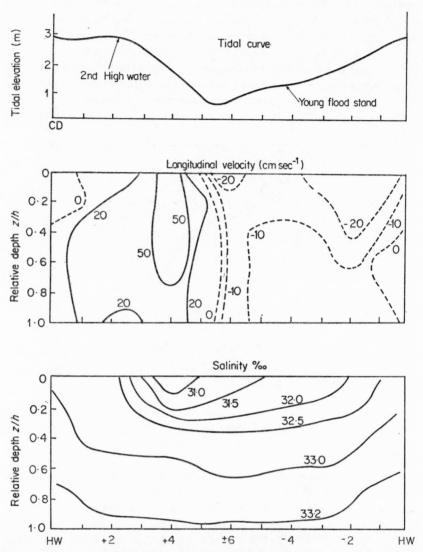

Figure 4.26   Southampton Water. Water elevation, velocity and salinity variation over a tidal cycle, Station 5, 9th August, 1966

surface current reverses up to 45 minutes after low water in the upper reaches. The surface currents are also affected by the time delay along the estuary of low and high waters, which between Calshot and Southampton Docks can reach 10 minutes.

The mean salinites and velocities for Station 5 are shown in Figure 4.27. The mean salinities show a zone of high salinity gradient at about mid-depth separating two homogeneous water masses. Towards the mouth of the estuary the surface to bottom difference in mean salinities decreases, the mid-depth zone of high gradient lessens and the mean curve becomes almost linear. In the upper part of the estuary the stratification increases in magnitude and the surface layer loses its homogeneity, though the bottom water is still almost homogeneous. In the shallower sections of the estuary only the top half of the curve is seen and in these areas the saline water forms a thin layer on the bottom with salinity gradients of up to 1‰ per metre at certain stages of the tide.

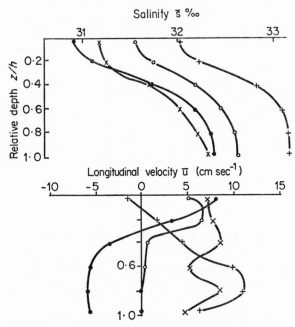

Figure 4.27 Southampton Water, mean salinity and velocity. ● 6th April, 1966, × 5th–6th May, 1966, ○ 24th–25th May, 1966, + 9th August, 1966

The mean velocities are typical of those in Southampton Water, showing considerable variation in the pattern of the mean velocity with depth. The expected pattern of a seaward surface flow overlying a landward bottom flow was only apparent on two occasions. At other stations there were similar variable results and for the seaward passage of river discharge considerable lateral variations in the mean flow are necessary. On the 9th August, 1966 the depth mean velocities at the four stations A1–4 across the channel were $+4\cdot28$, $-0\cdot18$, $-8\cdot50$ and $-10\cdot10$ cm sec^{-1} respectively. It is possible, however, that some of this variation may be errors accumulated in the measuring and averaging processes.

The distribution of mean salinity along the estuary for the different periods is shown in Figure 4.28. The effect of shallower water and the concentration of less saline water on the left-hand side is apparent at Stations 3, 4 and 6. The effect of the Power Station outfall in the region of Station 9 is most noticeable. The increase of salinity towards the mouth is almost linear and the gradients alter little at the periods of observation. The mean salinity distribution with distance inside the mouth can be represented by:

$$\langle \bar{s} \rangle = 32 \cdot 4 - 0 \cdot 12x \text{ where } x \text{ is in km.}$$

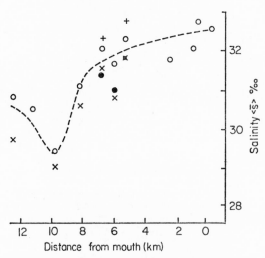

Figure 4.28 Southampton Water, longitudinal distribution of mean salinity. Symbols as in Figure 4.27

Determination of the linear regression of salinity against river discharge and tidal amplitude at Station 5 for the four periods gave:

$$\langle \bar{s} \rangle = 32 \cdot 77 - 0 \cdot 005T - 0 \cdot 00017R$$

where T is the tidal range in metres, R is the mean river discharge for the 7 days prior to the survey in m^3 sec^{-1} and $\langle \bar{s} \rangle$ is the depth averaged tidal mean salinity.

The lateral variations of tidal mean velocity make the estuarine classification of Hansen and Rattray difficult to apply. However, Southampton Water appears to be within the 2a category, a well mixed estuary with slight stratification in which the net flow reverses at depth and both advection and diffusion contribute to the upstream salt flux. This is similar to the Mersey Narrows.

The vertical velocities calculated for Sections A and B are shown on Figure 4.29. They indicate a tendency for upward flow on the western side of

Southampton Water and a downward flow on the shallower eastern side. The maximum vertical velocity was $0 \cdot 1$ cm sec^{-1} near the bottom at Station A1. This could be produced by the calculated mean longitudinal current acting on a slope of $1°10'$.

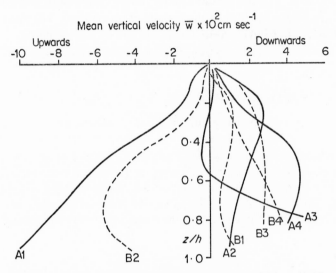

Figure 4.29 Southampton Water, mean vertical velocities calculated from measurements of 9th–10th August, 1966 at Sections A and B

There is some comparison between the water circulation in the Vellar at low river flow and in Southampton Water. It appears that when tidal mixing predominates the circulation of water is upwards away from the deeper water areas where the saline bottom water is thickest. This is in agreement with the required upward transport from the salt water into the fresher surface layer, and it is in contrast to the downward flow into the deeper areas that occurs when river flow predominates.

CHAPTER 5

Salt Balance

GENERAL FORMULATION

So far we have examined the distribution of salinity and the velocities in estuaries and we have made general statements about the principle mechanisms involved in mixing the salt and fresh water. In order to predict estuarine characteristics, it is necessary to be able to quantify the mixing processes. This is done by considering the budget of salt within sections of the estuary, by adding up the mass of salt being carried into a particular volume and equating it with what comes out, and the change of salinity within the volume.

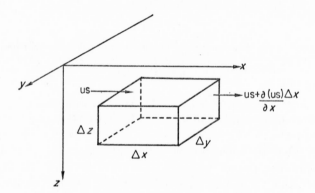

Figure 5.1   Salt transport through an elemental volume

The requirement for this treatment is that salt can be considered a conservative property and can be treated with a continuity equation similar to that which enabled us to calculate the mean vertical velocities. Thus an accurate knowledge of the distribution of salinity and velocity will be necessary. Though the results of the analysis will strictly be applicable only to the salt distribution, it will in fact give information on the probable distribution of both dissolved and suspended substances that can be considered to act in a similar way to the salt content.

Consider a small elemental volume of an estuary with sides of length Δx, Δy and Δz (Figure 5.1). Conservation of salt requires that the net inflow of

64

salt through the sides in a time Δt will equal the increase of salt content within the volume. The advective flow of salt through the face $\Delta y \Delta z$ in the time Δt is $us \Delta y \Delta z \Delta t$. The flow of salt through the opposite face in the same time, according to Taylors' series, is

$$us\Delta y \Delta z \Delta t + \frac{\partial (us)}{\partial x} \Delta x \Delta y \Delta z \Delta t$$

Consequently, the net inflow of salt in the x direction is

$$-\frac{\partial (us)}{\partial x} \Delta x \Delta y \Delta z \Delta t$$

Similarly in the y and z directions the inflow of salt is

$$-\frac{\partial (vs)}{\partial y} \Delta y \Delta x \Delta z \Delta t \text{ and } -\frac{\partial (ws)}{\partial z} \Delta z \Delta x \Delta y \Delta t$$

The molecular diffusion of salt through the face $\Delta y \Delta z$ will be

$$-\varepsilon \frac{\partial s}{\partial x} \Delta y \Delta z \Delta t$$

and through the opposite face it will be

$$-\varepsilon \frac{\partial s}{\partial x} \Delta y \Delta z \Delta t - \frac{\partial}{\partial x} \left(\varepsilon \frac{\partial s}{\partial x} \right) \Delta x \Delta y \Delta z \Delta t$$

where ε is the coefficient of molecular diffusion. The net diffusion in the x direction will thus be

$$+\frac{\partial}{\partial x} \left(\varepsilon \frac{\partial s}{\partial x} \right) \Delta x \Delta y \Delta z \Delta t = +\varepsilon \frac{\partial^2 s}{\partial x^2} \Delta x \Delta y \Delta z \Delta t$$

Similarly the diffusion in the other directions will be

$$+\varepsilon \frac{\partial^2 s}{\partial y^2} \Delta y \Delta x \Delta z \Delta t$$

and

$$+\varepsilon \frac{\partial^2 s}{\partial z^2} \Delta z \Delta x \Delta y \Delta t$$

The amount of salt present in the volume at the time t is $s\Delta x \Delta y \Delta z$. At the time $t + \Delta t$, according to Taylors' series, the amount present will be

$$s\Delta x \Delta y \Delta z + \frac{\partial}{\partial t} (s \Delta x \Delta y \Delta z) \Delta t$$

The net increase in salt is

$$\frac{\partial s}{\partial t} \Delta x \Delta y \Delta z \Delta t$$

Thus, in the absence of any creation of salt

$$\frac{\partial s}{\partial t} = -\frac{\partial(us)}{\partial x} - \frac{\partial(vs)}{\partial y} - \frac{\partial(ws)}{\partial z} + \varepsilon\left(\frac{\partial^2 s}{\partial x^2} + \frac{\partial^2 s}{\partial y^2} + \frac{\partial^2 s}{\partial z^2}\right) \qquad (5.1)$$

This is the equation of continuity for salt for instantaneous values. The continuity equation for volume can be developed in the same way and, considering the fluid to be incompressible, leads to equation (3.16). In estuaries the instantaneous salinity and velocities can be separated into a tidal mean, a fluctuation of tidal period and a short period turbulent fluctuation. Thus: $s = \bar{s} + S + s'$, $u = \bar{u} + U + u'$, $v = \bar{v} + V + v'$ and $w = \bar{w} + W + w'$.

The left-hand side if averaged over a tidal cycle becomes $\partial\bar{s}/\partial t$ because the tidal fluctuation S will average zero over a tidal cycle, and s′ by definition must be zero.

The first term on the right-hand side when multiplied out gives:

$$-\frac{\partial}{\partial x}(\bar{u}\bar{s} + \bar{u}S + \bar{u}s' + U\bar{s} + US + Us' + u'\bar{s} + u'S + u's')$$
$$\quad\;\; 1 \qquad 2 \qquad 3 \qquad 4 \qquad 5 \qquad 6 \qquad 7 \qquad 8 \qquad 9$$

and there is a similar set of terms for lateral and vertical components. Each of these terms can now be averaged over a tidal cycle.

Term 1. The average of this over a tidal cycle is $\bar{u}\bar{s}$.

Terms 2 and 4. There appears to be no reason why the tidal salinity variation should be correlated with the mean velocity, or vice versa. Consequently the tidal mean of these terms is considered negligible.

Terms 6 and 8. As we have already seen, the amplitude of velocity fluctuations increases with tidal current. Consequently there is a possibility that a correlation may exist between the salinity fluctuations and the tidal velocities, and between the velocity fluctuations and the tidal variation of salinity. If we can assume that these correlations do not change over the section of the estuary under consideration, then these terms also will be negligible.

Terms 3 and 7. It is unlikely that the salinity fluctuations will be correlated with the mean velocity and the velocity fluctuations with the mean salinity.

Term 5. If tidal velocity and salinity were both simple trigonometrical functions with a 90° phase difference between them, then the average of their cross-products would be zero. This may be valid in many estuaries for, though $\overline{US}$ may be large, its longitudinal variation $\partial/\partial x(\overline{US})$ may be small. This means that the correlation between the tidal fluctuations of salinity and velocity does not change significantly in the section of the estuary considered. This is most likely to occur in the lower parts of the estuary where the longitudinal salinity gradients are almost linear. In the upper reaches of the estuary, where the mean salinity falls off exponentially with distance and where the tidal velocities become markedly disturbed by the river, then $\partial/\partial x(\overline{US})$ may not be negligible.

It is possible to assess the contribution of this term because the observed salinity s_o and velocity u_o can be defined as

$$s_o = \bar{s} + S \quad \text{and} \quad u_o = \bar{u} + U$$

Consequently

$$\overline{US} = \overline{(s_o - \bar{s})(u_o - \bar{u})} \tag{5.2}$$

where the bar denotes averaging over a tidal cycle.

It is apparent now that if the values are observed over the wrong length of time then some of the shorter period turbulent fluctuation will appear in the tidal fluctuation cross-product. As the correlations between turbulent fluctuations of salinity and velocity are likely to change rapidly in the longitudinal direction, then significant values for $\delta/\delta x(\overline{US})$ can be produced. Consequently in a general treatment it may be advisable to retain this term. Similar arguments concerning the importance of the time averages of the various lateral and vertical terms can also be used. However, neglecting $\partial/\partial y(\overline{VS})$ and $\partial/\partial z(\overline{WS})$ may again not be valid.

Term 9. If turbulent fluctuations of velocity and salinity are correlated this term will be appreciable even when averaged over a tidal cycle.

Many analyses do neglect the tidal fluctuation cross-products and the salt balance for values averaged over a tidal cycle and neglecting molecular diffusion, will then be given by:

$$\frac{\partial \bar{s}}{\partial t} = -\frac{\partial(\bar{u}\bar{s})}{\partial x} - \frac{\partial(\bar{v}\bar{s})}{\partial y} - \frac{\partial(\bar{w}\bar{s})}{dz} - \frac{\partial\overline{(u's')}}{\partial x} - \frac{\partial\overline{(v's')}}{\partial y} - \frac{\partial\overline{(w's')}}{\partial z} \tag{5.3}$$

where the bar denotes averaging over a tidal cycle.

In this equation, the equation of salt continuity, the first three terms on the right-hand side are the advection terms, the salt flux caused by the mean flow, and the last three are the eddy–diffusion terms, the salt flux caused by short period eddies, which will be much larger than the molecular diffusion. The advective terms involve a mass flux of water as well as salt, whereas the diffusion terms are only associated with a flux of salt.

In general there is an approximate equilibrium between the terms on the right-hand side of equation (5.3), so that there is a slow time change of salt content. Under steady-state conditions $\partial \bar{s}/\partial t$ will be zero, and the advective and diffusive terms will balance.

Though it is possible to measure the contribution of the advection terms, the eddy–diffusion terms are all unknown. Consequently considerable modification to equation (5.3) is necessary before it can be used in a realistic situation. The most common modification is made by assuming either that there are no lateral variations, or by dealing with width averaged values of salinity and velocity. In this case the equation reduces to four terms and solution may be possible in certain circumstances. Other estuaries may be considered to be sectionally homogeneous, in which case the vertical diffusion

can be neglected and horizontal diffusion will be the only term to be calculated. However, this is true of very few, if any, estuaries.

The eddy–diffusion terms are generally rewritten in a form analogous to
that for molecular diffusion, i.e. a constant times the salinity gradient.

$$\overline{(u's')} = -K_x\frac{\partial \bar{s}}{\partial x}, \ \overline{(v's')} = -K_y\frac{\partial \bar{s}}{\partial y} \text{ and } \overline{(w's')} = -K_z\frac{\partial \bar{s}}{\partial z} \qquad (5.4)$$

Thus, the general case for three dimensions (equation 5.3) now becomes:

$$\frac{\partial \bar{s}}{\partial t} = -\frac{\partial (\bar{u}\bar{s})}{\partial x} - \frac{\partial (\bar{v}\bar{s})}{\partial y} - \frac{\partial (\bar{w}\bar{s})}{\partial z} + \frac{\partial}{\partial x}\left(K_x\frac{\partial \bar{s}}{\partial x}\right) + \frac{\partial}{\partial y}\left(K_y\frac{\partial \bar{s}}{\partial y}\right) + \frac{\partial}{\partial z}\left(K_z\frac{\partial \bar{s}}{\partial z}\right) \qquad (5.5)$$

This is the classical, Fickian, form of the equation of continuity for salt. The
coefficients are called the coefficients of eddy–diffusion and have the dimensions of length squared divided by time. These coefficients represent the
mixing conditions averaged over a tidal cycle and will have a different
physical meaning from the coefficients which would result from using a
shorter averaging time of, say, a minute or so. Bowden has termed the coefficients relating to the tidal mean values, effective eddy–diffusion coefficients.

It is also possible, by using the continuity equation for water (equation 3.16)
to write equation (5.5) as

$$\frac{\partial \bar{s}}{\partial t} = \bar{u}\frac{\partial \bar{s}}{\partial x} + \bar{v}\frac{\partial \bar{s}}{\partial y} + \bar{w}\frac{\partial \bar{s}}{\partial z} - \frac{\partial}{\partial x}\left(K_x\frac{\partial \bar{s}}{\partial x}\right) - \frac{\partial}{\partial y}\left(K_y\frac{\partial \bar{s}}{\partial y}\right) - \frac{\partial}{\partial z}\left(K_z\frac{\partial \bar{s}}{\partial z}\right) \qquad (5.6)$$

Certain modifications to this general equation are possible if the velocities
and salinities can be considered uniform in one or two of the axial directions.
These modifications have been developed by Pritchard (1958).

For a stratified estuary, an estuary with no spatial variation of salinity in
the *y* direction, but with variations in the *x* and *z* directions, the continuity
equation for volume is:

$$\frac{\partial \bar{b}\bar{u}}{\partial x} + \frac{\partial \bar{b}\bar{w}}{\partial z} = 0 \qquad (5.7)$$

where $\bar{b}$ is the mean estuary breadth. The equation for salt continuity is:

$$\frac{\partial (\bar{b}\bar{s})}{\partial t} = -\frac{\partial (\bar{b}\bar{u}\bar{s})}{\partial x} - \frac{\partial (\bar{b}\bar{w}\bar{s})}{\partial z} + \frac{\partial}{\partial x}\left(K_x\bar{b}\frac{\partial \bar{s}}{\partial x}\right) + \frac{\partial}{\partial z}\left(K_z\bar{b}\frac{\partial \bar{s}}{\partial z}\right) \qquad (5.8)$$

Combining equations (5.7) and (5.8) gives

$$\frac{\partial \bar{s}}{\partial t} = -\bar{u}\frac{\partial \bar{s}}{\partial x} - \bar{w}\frac{\partial \bar{s}}{\partial z} + \frac{1}{\bar{b}}\frac{\partial}{\partial x}\left(K_x\bar{b}\frac{\partial \bar{s}}{\partial x}\right) + \frac{1}{\bar{b}}\frac{\partial}{\partial z}\left(K_z\bar{b}\frac{\partial \bar{s}}{\partial z}\right) \qquad (5.9)$$

For a vertically homogeneous estuary with lateral variation:

$$\frac{\partial}{\partial x}(\bar{h}\bar{u}) + \frac{\partial (\bar{h}\bar{v})}{\partial y} + \frac{\partial \bar{h}}{\partial t} = 0 \qquad (5.10)$$

and

$$\frac{\partial(\bar{h}\bar{s})}{\partial t} = -\frac{\partial(\bar{h}\bar{u}\bar{s})}{\partial x} - \frac{\partial(\bar{h}\bar{v}\bar{s})}{\partial y} + \frac{\partial}{\partial x}\left(K_x\bar{h}\frac{\partial\bar{s}}{\partial x}\right) + \frac{\partial}{\partial y}\left(K_y\bar{h}\frac{\partial\bar{s}}{\partial y}\right)$$ (5.11)

Combining equations (5.10) and (5.11)

$$\frac{\partial\bar{s}}{\partial t} = -\bar{u}\frac{\partial\bar{s}}{\partial x} - \bar{v}\frac{\partial\bar{s}}{\partial y} + \frac{1}{\bar{h}}\frac{\partial}{\partial x}\left(K_x\bar{h}\frac{\partial\bar{s}}{\partial x}\right) + \frac{1}{\bar{h}}\frac{\partial}{\partial y}\left(K_y\bar{h}\frac{\partial\bar{s}}{\partial y}\right)$$ (5.12)

where $\bar{h}$ is the mean depth.

For a sectionally homogeneous (one-dimensional) estuary

$$\frac{\partial}{\partial x}(\bar{A}\bar{u}) + \frac{\partial\bar{A}}{\partial t} = 0$$ (5.13)

and

$$\frac{\partial(\bar{A}\bar{s})}{\partial t} = -\frac{\partial}{\partial x}(\bar{A}\bar{u}\bar{s}) + \frac{\partial}{\partial x}\left(\bar{A}K_x\frac{\partial\bar{s}}{\partial x}\right)$$ (5.14)

Combining equations (5.13) and (5.14)

$$\frac{\partial\bar{s}}{\partial t} = -\bar{u}\frac{\partial\bar{s}}{\partial x} + \frac{1}{\bar{A}}\frac{\partial}{\partial x}\left(\bar{A}K_x\frac{\partial\bar{s}}{\partial x}\right)$$ (5.15)

where $\bar{A}$ is the mean cross-sectional area.

The neglect in equations (5.11) and (5.14) of the tidal fluctuation cross-products may lead to large errors. In these cases, though $\partial/\partial x(\overline{US})$ may be small, the presence of a progressive element in the tidal wave may make

$$\frac{\partial}{\partial x}(\overline{hUS}) \quad \text{and} \quad \frac{\partial}{\partial x}(\overline{AUS})$$

significant, where $A = A_0 \cos(\omega t + \theta)$, (see p. 30).

Application of equations (5.7)–(5.15) to estuaries which are not strictly homogeneous in the ways defined, by using width, depth or cross-sectionally averaged values of salinity and velocity, will alter the meaning of the eddy–diffusion coefficients. For instance, if equation (5.12) is used in an estuary where there is in fact stratification, the coefficients K_x and K_y would not represent the mean value of the diffusion coefficients over the vertical. They would become functions which would allow the equation to describe the distribution of the mean salt content, and K_x and K_y would not have the physical significance ascribed to them in equation (5.4) (Pritchard, 1958).

For the different estuarine types, though, it can be expected that different terms in equation (5.5) will predominate, other terms being negligible. This is discussed by Pritchard (1955) who reasons that the following salt balances should hold in steady-state conditions.

Salt wedge estuary. The balance is between longitudinal and vertical advection, with the vertical diffusive flux important in the upper layer.

$$0 = \bar{u}\frac{\partial\bar{s}}{\partial x} + \bar{w}\frac{\partial\bar{s}}{\partial z}$$ (5.16)

Partially mixed. With increasing turbulent mixing the vertical diffusion term becomes important throughout the water column.

$$0 = \bar{u}\frac{\partial \bar{s}}{\partial x} + \bar{w}\frac{\partial \bar{s}}{\partial z} - \frac{\partial}{\partial z}\left(K_z\frac{\partial \bar{s}}{\partial z}\right) \tag{5.17}$$

Vertically homogenous with lateral variations. The vertical terms are now unimportant and lateral advection and diffusion are large.

$$0 = \bar{u}\frac{\partial \bar{s}}{\partial x} + \bar{v}\frac{\partial \bar{s}}{\partial y} - \frac{\partial}{\partial y}\left(K_y\frac{\partial \bar{s}}{\partial y}\right) \tag{5.18}$$

Sectionally homogeneous. The balance is entirely between longitudinal advection and diffusion.

$$0 = \bar{u}\frac{\partial \bar{s}}{\partial x} - \frac{\partial}{\partial x}\left(K_x\frac{\partial \bar{s}}{\partial x}\right) \tag{5.19}$$

In *fjords*, the mixing characteristics will depend largely on the depth of the sill at the mouth. For a deep sill the mixing characteristics are likely to be similar to the partially mixed type, but with the bottom layer replaced with a basin of undiluted sea water. For the shallower sill depth the mixing is more likely to be similar to the salt wedge estuary.

We shall now consider some examples of the applications that have been made of the equation of salt continuity. This should illustrate the validity of using reduced forms of equation (5.5) in different types of estuary.

KNUDSENS' HYDROGRAPHICAL THEOREM

If we can consider the mean flow as occurring in two layers and if we can assume that diffusion is negligible, then Knudsens' Hydrographical Theorem can be developed from consideration of salt and volume continuity under steady-state conditions.

If A denotes the cross-sectional area of each layer in Figure 5.2 then volume continuity requires that

$$A_1u_1 - A_3u_3 = A_2u_2 - A_4u_4$$

and continuity of salt requires that

$$A_1u_1S_1 - A_3u_3S_3 - A_2u_2S_2 + A_4u_4S_4 = 0$$

For the steady-state the net rate of transport of salt across each section is zero then

$$A_1u_1S_1 = A_3u_3S_3 \quad \text{and} \quad A_2u_2S_2 = A_4u_4S_4$$

Thus

$$A_2u_2\left(1 - \frac{S_2}{S_4}\right) = A_1u_1\left(1 - \frac{S_1}{S_3}\right)$$

and

$$A_4 u_4 \left(\frac{S_4}{S_2} - 1\right) = A_3 u_3 \left(\frac{S_3}{S_1} - 1\right)$$

If the section comprising A_1 and A_3 is the head of the estuary and R is the river flow:

$$A_1 u_1 - A_3 u_3 = R \quad \text{and} \quad A_1 u_1 S_1 = A_3 u_3 S_3 = 0$$

Consequently $R = A_2 u_2 - A_4 u_4$ and $A_2 u_2 S_2 = A_4 u_4 S_4$

Thus

$$A_2 u_2 = \frac{R S_4}{S_4 - S_2} \quad \text{and} \quad A_4 u_4 = \frac{R S_2}{S_4 - S_2} \tag{5.20}$$

Between the two sections vertical advection completes the circuit and

$$\overline{w} B = A_4 u_4 = R S_4 / S_4 - S_2 \tag{5.21}$$

where $\overline{w}$ is the mean vertical velocity and B is the surface area of the interface between the two sections.

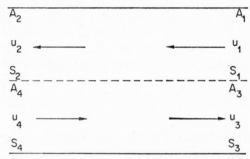

Figure 5.2 Longitudinal section of an estuary showing notation for Knudsen's Hydrographical Theorem

This approach is only valid for a purely advective salt balance and consequently may only be valid in salt wedge estuaries and fjords. The Hydrographical theorem is not a different concept, but only a formulation of a special case of the generalized salt balance equation.

HIGHLY STRATIFIED ESTUARIES

Vellar Estuary

In chapter 4 we have considered the distribution of salinity and current velocity in the Vellar. During the surveys measurements were taken at six stations, three on each of two sections, with measurements taken over a complete tidal cycle. The calculated mean vertical velocities have already been discussed. For calculation of the salt balance the gradients of the advective terms were obtained from the mean values at adjacent stations on each section, between the two sections, and for each 0·5 m depth. From this the advective flux of salt through each face of a unit volume was calculated. Table 5.1 shows some of the results for the survey of 15th February, 1967 at

TABLE 5.1

Salt Balance in the Vellar Estuary, Stations 5–7, 15th February, 1967 [a]

Station	Depth (m)	$\dfrac{\partial(\bar{u}\bar{s})}{\partial x}$	$\dfrac{\partial(\bar{v}\bar{s})}{\partial y}$	$\dfrac{\partial(\bar{w}\bar{s})}{\partial z}$	$\dfrac{\partial \bar{s}}{\partial t}$	$\dfrac{\partial(\overline{US})}{\partial x}$	Residual flux
		$\times 10^5$‰ per sec					
5	0·5	−1	−284	372	1	−1	87
	1·0	15	52	−6	0	25	86
	1·5	−13	202	−116	4	−3	74
6	0·5	51	−54	7	2	−5	1
	1·0	−2	−88	93	2	−4	1
	1·5	−4	0	25	3	1	25
	2·0	−12	147	−19	4	−1	119
7	0·5	41	202	−222	2	−13	10
	1·0	−14	−153	78	0	−6	−95
	1·5	−46	−168	51	1	−1	−163
	2·0	−178	230	−50	0	4	−4

[a] Reproduced with permission from K. R. Dyer and K. Ramamoorthy, *Limnol. Oceanog.*, **14**, 4–15, 1969, Table 1.

a low river flow stage. Of course, the errors inherent in this sort of analysis are large and it is estimated that, for that set of data, the errors in the longitudinal, lateral and vertical advective terms are about 5, 20 and 50×10^{-5}‰ sec^{-1} respectively. The time rate of change of salinity was estimated from the net change in salinity over the tidal cycle. It had a maximum value of $1·5 \times 10^{-4}$‰ sec^{-1} during the survey at high river flow.

The contribution of the tidal fluctuation term $\partial/\partial x(\overline{US})$ was calculated using equation (5.2). The calculated values of $\overline{(US)}$ were large, but the longitudinal variation was generally small, as can be seen from Table 5.1. This means that for periods greater than 1 minute or so the correlation between fluctuations of salinity and velocity did not change significantly in the section of the estuary considered. Consequently, a minute appears to have been a reasonable averaging time for the short period turbulent fluctuations. However, because of the small averaging period and the errors in individual measurements of both velocity and salinity, it is difficult to assess this term accurately. By analogy, the terms $\partial/\partial y(\overline{VS})$ and $\partial/\partial z(\overline{WS})$ were considered also small.

Because of the impossibility of estimating the separate contributions of the diffusion terms, the advective terms were evaluated and the residual flux was considered as being the result of eddy–diffusion in an indeterminate direction. The sign of the residual flux, however, indicates whether salt is being abstracted from or replaced into the volume under consideration by turbulent diffusive processes. The residual flux will also include the accumulated errors and the contributions of terms otherwise neglected.

Generally the advective terms did not balance, leaving quite a large proportion of the salt balance to be effected by diffusive processes. On 15th February the net flux carried by eddy diffusion averaged 51 % of the flux due to the largest advective term and in 79 % of the segments considered salt was being diffused into the volume to make up for a deficit in advection. The diffusive contribution was greatest near the halocline and diffusion was transporting salt out of the deeper, more saline water on the right-hand side and into the shallower fresher water on the left, a result which agrees with the mixing pattern one would expect.

The data for the period of flood, 20th January, were treated similarly, though they were not as complete. On this occasion the net eddy–diffusion flux averaged $57 \cdot 2 \%$ of the largest advective term, with a maximum value of about $600 \times 10^{-5}\%_{00}$ sec^{-1}. In 75 % of the segments, salt was being abstracted by diffusion. At this period in the fresh surface water there was no diffusion of salt, but large values occurred at the intense halocline near the bottom.

At the intermediate period, 9th February, the net flux by eddy–diffusion averaged 35 % of the largest advective term with a maximum value of about $900 \times 10^{-5}\%_{00}$ sec^{-1}. In 56 % of the observations, salt was being diffused into the volume under consideration.

Thus it seems that, with decreasing river discharge, the proportion of the salt flux effected by eddy–diffusion passed through a minimum and the sense of the diffusion changed. At low river flow salt was mainly diffused into the volume and at high river flow diffusion abstracted salt. It is possible that these changes may be related to the development of a landward residual saline bottom flow, the reversal in the circulation of the secondary flow system and the increased importance of tidal mixing with respect to river flow.

FJORDS

Silver Bay

The salt balance at the entrance of Silver Bay, Alaska, has been investigated by McAlister *et al.* (1959). As described in chapter 4 measurements were obtained on three sections with high river discharge in July 1956 and with low discharge in March 1957. Current measurements were obtained only at the centre section, but salinity profiles were observed on all three sections. Lateral variations were small so the observed results were averaged across the width.

For the current measurements, the results from the centre station on the middle section were doubly weighted to represent better the cross-sectional mean.

The form of the mean salinity profiles was similar at all sections. The velocity profiles were also assumed to have the same form at the three sections, so that the velocities at any depth had the same ratio to the surface velocity.

In order to obtain the gradient of the mean longitudinal velocities, it was assumed that all the salt added to the surface layer between the sections came from below by upward vertical advection. Consequently, water of salinity $23 \cdot 6\%_0$ entering the upper section in the high river flow stage was supplemented by water of salinity $31 \cdot 2\%_0$ from below. As the salt transport through the centre section was known, the salt transport and consequently the mean velocity profiles at the other sections were calculated. This process also permitted calculation of the vertical velocities.

Horizontal diffusion was considered in a similar way to Pritchard's analysis of the James River (Pritchard, 1954). A section of the fjord which was bounded by the two cross-sections 1 and 3, and by the bottom, sides and surface of the fjord was considered. Within this section the salt balance, for a steady-state situation, must be entirely provided by horizontal advection and diffusion and can be written as:

$$\iint_\sigma \bar{u}\bar{s} \, d\sigma_x + \iint_\sigma (\overline{u's'}) \, d\sigma_x = 0 \qquad (5.22)$$

where $\iint_\sigma d\sigma_x$ represents the integral over the bounding surfaces of the segment. The first term of equation (5.22) was considered in two parts, a positive part in the seaward flowing surface layer and a negative part in the inflowing lower layer. These two parts almost balanced, leaving only a small amount of salt to be brought into the section by turbulent diffusion. The second, diffusion, term was less than 10% of either of the two parts of the advective term, for both high and low run-off conditions, and horizontal diffusion was therefore considered negligible. This is not strictly valid since the diffusive term is the mean value over the area of a cross-section and may be the result of very unequal contributions in the two layers. As the tidal variation of velocity is small compared with the mean values, omission of the term $\iint_\sigma \overline{US} d\sigma_x$ from equation (5.22) is reasonable.

The salt balance was then calculated for segments bounded by the two cross-sections and between the bottom and a height H above the bottom. The salt balance equation for no local change in salt content then is:

$$0 = \iint_{\sigma_H} \bar{u}\bar{s} d\sigma_x - (\overline{w's'})\sigma_H - (\overline{w}\bar{s})\sigma_H \qquad (5.23)$$

The horizontal advective term was evaluated over the two sections at the ends of the segment and the vertical advective term over the top surface of the segment. Summation started at the bottom and segments were added progressively on top. The horizontal and vertical advections balance with small amounts left as the contribution of vertical diffusion. Since an entirely advective balance was assumed to enable calculation of the longitudinal variation in velocity, this result may represent the accumulated computational errors.

Hardanger Fjord

Saelen (1967) has used Knudsen's equations in analysing results from Hardanger Fjord. The main difficulty of the approach is in estimating the depth of water involved in the ingoing bottom flow near the mouth, without knowledge of the mean current velocity profile. It was estimated that the surface outflow at the mouth was between two to six times the river flow. Using equation (5.21) the mean vertical velocity was calculated as $2\text{--}8 \times 10^{-4}$ cm sec^{-1}, corresponding to a daily vertical movement of about half-a-metre.

PARTIALLY MIXED ESTUARIES

Southampton Water

Data from Southampton Water have been analysed in the same way as data from the Vellar. In this case four stations were established on each of two sections. The stations on each section were occupied in sequence throughout a complete tidal cycle. The sections were observed on subsequent days. In all cases the three advective terms almost cancelled leaving only a small proportion of the salt balance to be effected by diffusion (Table 5.2). The

TABLE 5.2

Salt Balance for Two Stations in Southampton Water, August 1966

Station	Depth (m)	$\dfrac{\partial(\overline{us})}{\partial x}$	$\dfrac{\partial(\overline{vs})}{\partial y}$	$\dfrac{\partial(\overline{ws})}{\partial z}$	$\dfrac{\partial(\overline{US})}{\partial x}$	Residual flux
		$\times 10^4\%_0$ per sec				
A1	0·66	−13·5	48·8	−38·7	0·0	−3·4
	2·60	1·2	13·5	−14·9	−0·3	−0·5
	5·25	19·0	9·5	−33·0	−0·1	−4·6
	7·85	31·0	0·0	−31·8	−0·1	−0·9
	10·50	38·7	−15·5	−24·5	+0·1	−1·2
	12·30	28·9	0·0	−29·1	0·0	−0·2
A4	0·90	−34·1	−9·8	44·5	0·0	0·6
	2·30	−24·6	−54·6	76·2	0·4	−2·6
	2·70	−17·6	35·2	−16·7	−0·1	0·8

calculated range of residual salt flux by eddy–diffusion had a magnitude of $11\cdot5\%\pm3\cdot5\%$ of the longitudinal advective term and in 64% of the segments salt was being added to the volume being considered, by turbulent processes. Maximum diffusion amounts occurred at mid-depth. Salt was apparently diffusing into the fresher water on the left-hand side of the estuary and out of the saltier water on the right hand-side.

The sense of this diffusion is the same as that from the Vellar at low river flow and agrees with the processes postulated for partially mixed estuaries. When the saline inflow is well developed, the amount of salt advected longitudinally landwards in the bottom layer on the mean flow decreases towards the head of the estuary. The surplus salt is abstracted by turbulent diffusion and added to the seaward surface flow giving increased longitudinal advection towards the mouth.

James River

Pritchard and Kent (1953) and Pritchard (1954) have considered the salt balance in the James River, a tributary of Chesapeake Bay. The estuary was assumed to be laterally homogeneous and comprised a series of elements of equal thickness, but whose width b varied with depth and along the estuary between the cross-sections 11, 17, 24.

Equations (5.7) and (5.9) for continuity of water and of salt were used in this case. The tidal fluctuation terms were considered negligible. The time change of mean salinity was also small. The vertical velocities were calculated using equation (5.7) and the two advective terms were then calculated from the observed data. The cross-sectional mean value of the horizontal diffusive flux $(\overline{u's'})$ was calculated using equation (5.22), the first term of which was again considered in two parts. The value of the second, diffusive, term was in all cases less than 5% of either of the two parts of the first term. The value of $(\overline{u's'})$ was assumed constant over each of the cross-sections and, from the results, plots of $\overline{b}(\overline{u's'})$ against x were constructed for different depths. This enabled calculation of the values at the different depths of

$$\frac{\partial}{\partial x}\overline{b}(\overline{u's'})$$

The resulting values for the horizontal diffusion term are small (Table 5.3).

Consequently the only unknown term in equation (5.9) is the vertical non-advective term, the last on the right-hand side. The values for this are found by simple addition and the results are given in Table 5.3. Of course, computational errors will also be accumulated in the vertical diffusive term.

The horizontal advective term and the vertical non-advective term are the most important. The vertical advective term becomes important at mid-depth near the halocline. In spite of the possible errors involved in the

TABLE 5.3

Salt Balance in the James River [a]

Depth (m)	$\partial \bar{s}/\partial t$	$\bar{u}\partial \bar{s}/\partial x$	$\bar{w}\partial \bar{s}/\partial z$	$1/b \, \partial/\partial x \, (b(\overline{u's'}))$	$1/b \, \partial/\partial z \, (b(\overline{w's'}))$
			$\times 10^4$ g m^{-3} s^{-1}		
0·0	−4·2	484·0	0·0	1·3	−481·1
0·5	−3·6	407·0	−1·0	1·3	−403·7
1·0	−1·0	326·0	−1·4	1·3	−324·6
1·5	0·0	238·0	−1·9	0·6	−236·7
2·0	−0·1	159·0	−4·2	0·2	−154·9
2·5	0·1	88·0	−21·9	−1·0	−65·2
3·0	0·8	−8·0	−117·0	−0·8	125·0
3·5	2·5	−118·0	−154·8	−0·6	270·9
4·0	4·8	−219·0	−81·0	−0·6	295·8
4·5	15·5	−279·0	−33·8	−0·6	297·9
5·0	14·5	−288·0	−13·7	−0·4	287·6
5·5	10·0	−278·0	−9·2	−0·7	277·9
6·0	10·5	−278·0	−6·7	−1·2	275·4
6·5	11·9	−286·0	−5·2	−0·6	279·9
7·0	12·0	−318·0	3·5	−1·0	303·5
7·5	12·0	−345·0	3·5	−0·1	329·6

[a] Reproduced with permission from D. W. Pritchard, *Jour. Marine Res.*, **13**, 1954, Table 1.

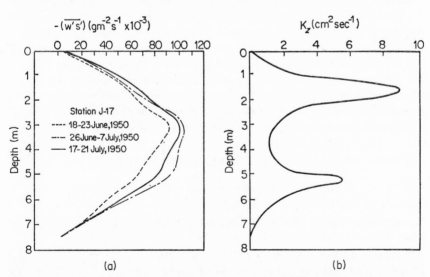

Figure 5.3   Station J17. (a), vertical eddy flux of salt $(\overline{w's'})$ as a function of depth; (b), vertical eddy diffusivity (K_z) as a function of depth. (Reproduced with permission from D. W. Pritchard and R. E. Kent, 1953, *Tech. Rept.* VI, Chesapeake Bay Inst., Johns Hopkins University, Figure 7 and D. W. Pritchard, 1967, in *Estuaries* (ed. G. H. Lauff) AAAS. Pub. No. 83, Figure 7. (Copyright 1967 by the American Association for the Advancement of Science)

computation of the values for horizontal diffusion the balance seems intuit-
ively correct. The vertical distribution of salinity represents a balance between
vertical eddy–diffusion, the horizontal advection and the vertical advection.

Assuming zero values of eddy–diffusion at the surface, integration of the
vertical diffusion data shows that the vertical non-advective flux of salt was a
maximum at mid-depth (Figure 5.3a). The vertical eddy–diffusion coefficient
K_z calculated from these results reached 9 cm^2 sec^{-1} (Figure 5.3b). As tidal
movements are the most important factor governing the mixing processes
there should be a relationship between the tidal current velocities and the
vertical non-advective flux of salt. Pritchard demonstrates this for three
periods of observation at three sections in the James River, the salt flux
rising with the tidal velocities.

Mersey Narrows

If one assumes, following Pritchard's work in the James River, that horizontal advection and vertical eddy–diffusion are the dominant processes
affecting the distribution of salinity, and that this is applicable to observed
as well as mean values, then estimates of the vertical eddy–diffusion coefficient
can be made (Bowden 1960, 1963).

The salt balance equation becomes:

$$\frac{\partial s_o}{\partial t} + u_o \frac{\partial s_o}{\partial x} = \frac{\partial}{\partial z}\left(K_z \frac{\partial s_o}{\partial z}\right) \tag{5.24}$$

Continuity requires that

$$\frac{\partial \langle s \rangle}{\partial t} + \left\langle u \frac{\partial s}{\partial x} \right\rangle = 0$$

where the brackets $\langle \rangle$ denote a depth mean taken from the surface to the
bottom. The salinity at any depth can be considered as a depth mean and a
deviation, i.e. $s_o = \langle s \rangle + s_1$. Then in the case of steady flow, at any depth

$$\frac{\partial s_1}{\partial t} + u_o \frac{\partial s_o}{\partial x} - \left\langle u_o \frac{\partial s}{\partial x} \right\rangle = \frac{\partial}{\partial z}\left(K_z \frac{\partial s_o}{\partial z}\right) \tag{5.25}$$

With zero boundary conditions at the surface equation (5.25) can be
integrated stepwise from the surface to the bottom and the distribution of K_z
found. The equation can also be integrated over a tidal cycle, but the vertical
eddy–diffusion coefficient will not then bear close relationship to the shorter
term values, as it will represent the mixing conditions averaged over a tidal
period. Values of this effective coefficient given by Bowden (1963) and Bowden
and Sharaf el Din (1966a) are given in Table 5.4.

In an attempt to investigate the temporal variation of vertical eddy–diffusion
coefficient Bowden (1963) applied equation (5.24) to the hourly data, but,
because of measurement errors, consistent results were obtained for only
one period. These indicated that when the current was large the values of K_z

were 3–5 times the effective values over a tidal period. The maximum value was 155 cm² sec⁻¹ at mid-depth, 3 hours after high water. These values, however, were still smaller than those which one would expect in conditions of neutral stability.

In partially mixed estuaries there are large vertical variations in the amount of horizontal advection. Because of the two-layer nature of the mean flow, there will be an upstream salt flux near the bottom and a downstream salt flux on the surface. These effects produce gravitational convection or density current transport.

TABLE 5.4

Coefficient of Vertical Eddy–Diffusion K_z cm² sec⁻¹ at Various Points in the Mersey Narrows.[a] (For station positions see Figure 4.18)

z/h	Station number						
	C2	C2	E2	C2	C1	C2	C3
0·1	5·5	9	5·5	2·8	8	6	5
0·3	18	24	27	9	12	15	13
0·5	28	33	71	28	22	28	17
0·7	23	34	41	28	22	23	9
0·9	8	16	26	15	1	3	1

[a] Reproduced with permission from K. F. Bowden, *Int. J. Air Wat. Poll.*, **7**, 1963, Table 3 and K. F. Bowden and S. H. Sharaf el Din, *Geophy. Jour.* **10**, 1966, Table 2.

Bowden and Gilligan (1971) have examined the variation of density current transport across various sections in the Mersey. To remove the lateral variations of mean current due to topographic effects the axis of profiles of mean current from different stations in the estuary were reset so that the depth mean of the residual current $\langle \bar{u} \rangle$ was equivalent to the river discharge. Then the density current transport across the section in the lower layer is

$$-F = \int_{z_1}^{h} b\bar{u} \, dz$$

where both b and $\bar{u}$ are functions of depth and z_1 is the depth of zero velocity. In the upper layer

$$F + R = \int_{0}^{z_1} b\bar{u} \, dz$$

From measurements over 13 separate tidal periods at stations on one section in the middle of the Narrows, the density current transport F was compared with river flow and root mean square tidal current velocity and:

$$F = 54 + 0.88 \, R_{10} + 7.25 \, u_t$$

F is in m³ sec⁻¹, R_{10}, the river flow averaged over 10 days prior to the measurements, is in m³ sec⁻¹ and u_t is cm sec⁻¹. The dependence of F on R_{10} appears to be relatively slight.

Measurements at other sections in the estuary showed that the density current transport decreased towards both ends of the Narrows. Also the density current transport decreased with time at the inner end of the Narrows, during a period of increasing river discharge, presumably as the lower layer was pushed downstream.

WELL MIXED AND HOMOGENEOUS ESTUARIES

In one-dimensional estuaries, where the distribution of properties can be considered dependent on x only, considerable use has been made of equation (5.14) integrated once with respect to x

Thus

$$\bar{u}\bar{s} = K_x \frac{\partial \bar{s}}{\partial x} \qquad (5.26)$$

This states that the longitudinal advection of salt, downstream on the sectional mean flow, is balanced by an upstream horizontal diffusion. The sectional mean velocity is related to the river flow since $\bar{u} = R/\bar{A}$.

Therefore

$$K_x = R\bar{s} \bigg/ A\frac{\partial \bar{s}}{\partial x} \qquad (5.27)$$

Thus K_x can be calculated for any position in an estuary if the river flow, the cross-sectional area and the distribution of salinity along the estuary are known. Some values are shown in Table 5.5. It is noticeable how K_x decreases with river flow.

This approach takes no account of lateral or vertical variations in the mean flow, however. The effect of any gravitational circulation on the salt transport will appear in the diffusive term, in spite of not being the result of a turbulent exchange. Consequently the diffusion coefficient is modified in its meaning. In equation (5.26), as the upstream salt flux is proportional to the salinity gradient, K_x is a coefficient of longitudinal eddy–diffusion. However, in equation (5.27) the experimentally observed upstream salt flux may be largely due to convection effects and K_x is thus a coefficient of longitudinal dispersion, or a coefficient of effective longitudinal diffusivity.

In a truly homogeneous estuary, one in which a uniform density is present at all times, there will be no gravitational convection, but dispersion can still occur. In this case the dispersion is caused by velocity shear due to the friction on the sea bed. If one considers a patch of dye in a channel, the dye near the sea bed travels slower than the average and vertical diffusion will mix it completely throughout the water column. The dye near the surface travels faster than average, but will be mixed downwards. Consequently the dye spreads out as the patch moves along the channel, because of the variations of longitudinal advection with depth, coupled with vertical eddy–diffusion. The dispersion of dye in unidirectional flow in pipes has been studied by Taylor (1954) who showed that for a pipe of radius a,

$$K_x = 10{\cdot}1\,aU_* \tag{5.28}$$

where U_* is the friction velocity, $U_* = \sqrt{\tau_0/\rho}$, τ_0 being the shear stress at the wall.

For open trapezoidal channels Elder (1959) found that

$$K_x = 5{\cdot}93\,hU_* \tag{5.29}$$

where h is the depth of flow.

TABLE 5.5

Longitudinal Eddy–Diffusion Coefficient K_x for some Estuaries Calculated Using equation (5.27) [a]

	Vertical salinity difference ‰	K_x 10^5 cm^2 sec^{-1}
Severn (summer) Aust	0·3	17·4
Portishead	0·1	10·6
Severn (winter) Aust	0·5	30·9
Portishead	0·3	15·7
Thames (low flow)		
10 miles below London Bridge	1·0	5·3
25 miles below London Bridge	1·0	8·4
Mersey Narrows		
low flow	1·3	16·1
high flow	1·5	36·0
Southampton Water	1·2	15·8

[a] Reproduced with permission from K. F. Bowden, *Int. J. Air Wat. Poll.*, **7**, 1963, Table 1, with additions.

Experimental results for natural streams, however, show the observed K_x is an order higher than these theoretical estimates. Of course the theoretical estimates are for unidirectional flow, but Bowden (1965) has shown that, with certain assumptions, the K_x for alternating flow is about half that for unidirectional flow. For a tidal current of amplitude U_o the longitudinal dispersion coefficient

$$K_x = 0{\cdot}15\,U_o h \quad \text{(Bowden, 1963)} \tag{5.30}$$

Values of K_x calculated using equation (5.30) are between 10 and 100 times smaller than those derived using equation (5.27) (Table 5.5). The correspondence between theory and observation can be improved by assuming more complicated velocity profiles (Bowden, 1965) and by consideration of lateral velocity shear (Fischer, 1967). Secondary currents are also probably very important in producing the observed dispersion effects.

4

SHEAR EFFECT

Since the longitudinal dispersion is probably the result of a combination of velocity shear on the walls and gravitational convection, it is interesting to approach the salt balance problem by examining the 'shear effect'. This approach considers the contributions of deviations of velocity and salinity from their depth means, to the salt flux, and to the magnitude of the dispersion coefficient.

The effect of vertical shear has been examined by Bowden (1963). An estuary without lateral variations is considered. The instantaneous rate of transport of salt through a unit width of a section perpendicular to the mean flow is given by:

$$Q = \int_0^h us\, dz = \langle us \rangle \tag{5.31}$$

where h is the depth. At any depth let $u = u_0 + u'$ and $s = s_0 + s'$, where u_0 and s_0 are the observed values averaged over a number of minutes, and u' and s' are the irregular, turbulent variations.

u_0 and s_0 may also be written as

$$u_0 = \langle u \rangle + u_1 \qquad \text{and} \qquad s_0 = \langle s \rangle + s_1$$

where
$$\langle u \rangle = \frac{1}{h}\int_0^h u_0\, dz \qquad \text{and} \qquad \langle s \rangle = \frac{1}{h}\int_0^h s_0\, dz$$

Thus u_1 and s_1 are the deviations with depth from the mean values over the entire depth, $\langle u \rangle$ and $\langle s \rangle$.

Because of tidal fluctuations $\langle u \rangle$ and $\langle s \rangle$ will vary regularly over a tidal cycle. Consequently $\langle u \rangle = \bar{u} + U$ and $\langle s \rangle = \bar{s} + S$

where
$$\bar{u} = \frac{1}{T}\int_0^T \langle u \rangle\, dt \qquad \text{and} \qquad \bar{s} = \frac{1}{T}\int_0^T \langle s \rangle\, dt$$

so that $\bar{u}$ and $\bar{s}$ are the mean values of $\langle u \rangle$ and $\langle s \rangle$ over the tidal cycle. U and S could perhaps be represented by $U_0 \cos(\omega t + \theta)$ and $S_0 \sin(\omega t + \theta)$.

Consequently equation (5.31) becomes, averaged over a tidal cycle,

$$\bar{Q} = \bar{h}\bar{u}\bar{s} + \frac{1}{T}\int_0^T hUS\, dt + \frac{1}{T}\int_0^T h\langle u_1 s_1 \rangle\, dt + \frac{1}{T}\int_0^T h\langle u's' \rangle\, dt \tag{5.32}$$

or
$$\bar{Q} = \bar{h}\bar{u}\bar{s} + (\overline{hUS}) + \overline{h\langle u_1 s_1 \rangle} + \overline{h\langle u's' \rangle}$$
$$= Q_1 + Q_2 + Q_3 + Q_4$$

where the brackets $\langle \rangle$ denote the depth means of the products $u_1 s_1$ and $u' s'$ and $\bar{h}$ is the mean depth.

The first term Q_1 represents the contribution to the total salt flux by the mean flow caused by the river discharge since $\bar{u} = R/\bar{A}$. Q_2 arises from the variation in the depth mean velocity and salinity during the tidal period. If the tidal fluctuations are harmonic and 90° out of phase Q_2 should be zero providing the tide acts purely as a standing wave. It will provide a finite negative amount if the water flowing through the section on the flood has a higher salinity than that flowing seaward on the ebb. This could occur if the tide does not act as a purely standing wave. Q_3 arises from the variation of the velocity and salinity with depth, the shear effect, and will be finite if the two are correlated. Thus a positive (downstream) u_1 associated with a negative s_1, a lower salinity, will produce a finite Q_3. This could also be termed a gravitational convection mode and includes the effect of vertical diffusion. Q_4 represents the salt flux on the short period turbulence.

Equation (5.32) can be written in the form

$$\overline{Q} = \overline{h}\overline{u}\overline{s} - K_x \overline{h}\frac{\partial \overline{s}}{\partial x} \tag{5.33}$$

The coefficient of longitudinal dispersion calculated this way includes the effect of Q_2 and Q_3 as well as the short period turbulent term Q_4. Consequently this coefficient will alter with magnitude of the vertical eddy–diffusion. Bowden (1965) has shown that K_x calculated this way is inversely proportional to K_z so that a decrease in vertical diffusion due to, for instance, a stable density gradient, gives rise to an increase in the longitudinal dispersion.

From data obtained in the Mersey Narrows, Bowden (1963) has calculated the values of Q_1, Q_2 and Q_3. From the observed data Q_4 cannot be calculated, but it is assumed small compared with the other terms. The values for four periods are shown in Table 5.6. Q_1, representing the advection on the mean

TABLE 5.6

Transport of Salt in the Mersey Narrows on Four Occasions Calculated Using equation (5.32) [a]

$Q_1/h‰$ cm sec^{-1}	6·34	2·63	10·00	6·00
$Q_2/h‰$ cm sec^{-1}	−6·95	−0·44	−12·27	−14·40
$Q_3/h‰$ cm sec^{-1}	−2·36	−3·04	−3·22	−5·15
$(Q_1+Q_2+Q_3)/h‰$ cm sec^{-1}	−2·97	−0·85	−5·49	−13·65

[a] Reproduced with permission from K. F. Bowden, *Int. J. Air Wat. Poll.*, 7, 1963, Table 2.

flow, is always downstream, but is more than compensated by the upstream salt flux of Q_2 and Q_3. Thus there is a net upstream flux of salt, small at the second period, but large during the other three. This may be the result of one or several of the following causes:

1. Horizontal diffusion on the short period turbulence (Q_4) which has been neglected.

2. A time change in mean salinity, so that $\partial \overline{s}/\partial t \neq 0$, i.e. $\overline{Q} \neq 0$.

3. Lateral variations across the estuary especially in velocity, caused by secondary currents.

On a subsequent occasion when the results at three stations on a cross-section were analysed (Bowden and Sharaf el Din, 1966a), lateral variations in $(Q_1 + Q_2 + Q_3)/\bar{h}$ were apparent. Also the values were positive, requiring a net transport of salt seawards through the section. This implied a decreasing salinity in the upstream part of the estuary. There is an additional difficulty in that Bowden and Sharaf el Din found that the mean value of $\bar{u}$ over the cross-section was an order higher than the flow deduced from the river discharge ($\bar{u} = R/\bar{A}$). The calculated values for Q_1 were for the smaller river discharge value. This discrepancy may be caused by imprecision in the current measurements coupled with undiscovered lateral variations of velocity. It may also be the effect of the mass transport caused by the progressive nature of the tidal wave as discussed in chapter 3. The value of $\bar{Q}$ could thus be locally higher than shown in Table 5.6.

TRANSVERSE SHEAR

Because of the lateral variations in current velocity and salinity, transverse shear is likely to be an important factor in producing longitudinal dispersion and its importance in the salt balance in estuaries has been considered in several ways.

Okubo (1964) has considered rigorously the averaging processes and their validity when applied to the mass and pollutant balance on a cross-section of an estuary. The instantaneous salt balance equation can be written as:

$$\frac{\partial (as_A)}{\partial t} = -\frac{\partial}{\partial x}[As_A u_A + A(s_d u_d)_A] \tag{5.34}$$

where the instantaneous salinity and velocity is considered as being composed of a cross-sectional mean and a deviation from this mean, i.e. $s = s_A + s_d$ and $u = u_A + u_d$. The subscript A denotes averaging over the cross-sectional area. The instantaneous values are then considered as a tidal average and a short period fluctuation. Thus $A = \bar{A} + A'$, $s_A = \bar{s}_A + s'_A$, $u_A = \bar{u}_A + u'_A$ and $(s_d u_d)_A = \overline{(s_d u_d)_A} + (s_d u_d)'_A$.

Substituting into equation (5.34), averaging over a tidal cycle and eliminating terms leads to:

$$\frac{\partial}{\partial t}(\bar{A}\bar{s}_A) = -\frac{\partial}{\partial x}[\bar{s}_A(\bar{A}\bar{u}_A + \overline{A'u'_A}) + \bar{A}(\overline{s'_A u'_A} + \overline{(s_d u_d)_A})] \tag{5.35}$$

Hansen (1965) has used the same approach, but considered each component as the sum of a tidal mean, a tidal oscillation and a turbulent fluctuation, all averaged over the cross-section. Thus $A = \bar{A} + A + A'$, $u_A = \bar{u}_A + U_A + u'_A$, $s_A = \bar{s}_A + S_A + s'_A$ and $(u_d s_d)_A = \overline{(u_d s_d)_A} + (U_d S_d)_A + (u_d s_d)'_A$.

The mean salt flux over a tidal cycle through a cross-section is:

$$\bar{F}_s = \overline{Au_A s_A} + \overline{A(u_d s_d)_A} \tag{5.36}$$

Thus,

$$\bar{F}_s = \overline{(\bar{A}+A+A')(\bar{u}_A+U_A+u'_A)(\bar{s}_A+S_A+s'_A)} + \overline{(\bar{A}+A+A')((u_ds_d)_A +}$$
$$\overline{(U_dS_d)_A+(u_ds_d)'_A)}$$

The mean flux of water through the section during the tide is:

$$R = \overline{A\bar{u}_A} + \overline{AU_A} + \overline{A'u'_A}$$

This is comparable to equation (3.15).

Multiplying out and eliminating terms gives

$$\bar{F}_s = R\bar{s}_A + \overline{AU_AS_A} + \bar{u}_A\overline{AS_A} + \overline{AU_AS_A} + \bar{A}\overline{(u_ds_d)_A} + \overline{A(U_dS_d)_A} +$$
$$(\overline{Au'_As'_A} + \bar{u}_A\overline{A's'_A} + \overline{A'u'_As'_A} + \overline{A'(u_ds_d)'_A}) \qquad (5.37)$$

The terms on the right-hand side of equation (5.37) express the seaward salt flux associated with:

1. The mean river flow and salinity.
2. Correlation of tidal variations of the sectional mean salinity and current.
3. Correlation between tidal variations of cross-section and mean salinity.
4. Third-order correlation of tidal variations in mean salinity, velocity and cross-section.
5. Mean shear effect.
6. Covariance of shear effect and cross-section.
7. Correlation between short period fluctuations.

Terms 2 and 5 are similar to Q_2 and Q_3 of equation (5.32).

From harmonic analysis of observations from the Columbia River, Hansen (1965) produced equations showing the amplitudes and phases of $\bar{A}+A$, $\bar{s}_A+S_A$, $\bar{u}_A+U_A$ and $\overline{(u_ds_d)}+(U_dS_d)_A$. These equations enabled calculation of Terms 1–6 of equation (5.37). The change in salt storage above the section was assumed to be small and the results showed that, of the salt advected seawards in the mean river discharge, about 40% was balanced by covariance between fluctuations of tidal period (Term 2) and about 45% was balanced by shear effects (Term 5). The remaining 15% was considered to be caused by the short period fluctuations (Term 7) as the other terms were small and were comparable with the uncertainty in estimation of the major fluxes.

It is interesting to note that the effect of a progressive component on the tidal wave in the Columbia River produces a large value for $\overline{AU_A}$. Also the phase relationships between the tidal fluctuations of velocity and salinity are such that large values for $\overline{AU_AS_A}$ are produced. Thus U_A and S_A are not 90° out of phase. However, the mean product $\overline{AU_AS_A}$ is small. Consequently, the progressive component of the tidal response is cancelling out the effect that the tidal fluctuations of salinity and velocity produce by not being 90° out of phase.

It would be instructive to analyse the shear effect in such a way that it is possible to separate the effects of vertical shear and transverse shear. Their comparative importance may illustrate better the processes operating to cause a balance of salt in estuaries.

If u_2 is the deviation of the depth mean velocity $\langle u \rangle$ from a cross-sectional mean velocity u_A and u_A has a tidal fluctuation given by $u_A = \overline{u_A} + U_A$, then the instantaneous velocity u can be represented by:

$$u = \overline{u_A} + u_1 + u_2 + U_A + u'$$

where u_1 is the deviation of the observed values from the depth mean. The salinity variations can be represented in a similar way and the cross-product averaged over a tidal cycle will be:

$$\overline{us} = \bar{u}_A \bar{s}_A + \overline{u_1 s_1} + \overline{u_2 s_2} + \overline{U_A S_A'} + \overline{u's'}$$

If $\overline{u_A} = R/\overline{A}$, then integration over the cross-section will give the mean salt flux as:

$$\overline{F}_s = R\bar{s}_A + b \int_0^h \overline{u_1 s_1}\, dz + h \int_0^b \overline{u_2 s_2}\, db + A\overline{U_A S_A} + A(\overline{u's'})_A$$

In steady-state conditions this becomes

$$0 = R\bar{s}_A + b \int_0^h \overline{u_1 s_1}\, dz + h \int_0^b \overline{u_2 s_2}\, db + A\overline{U_A S_A} - AK_x \frac{\partial \bar{s}_A}{\partial x} \qquad (5.38)$$

$$\;\;\underset{1}{\phantom{R\bar{s}_A}}\quad\underset{2}{}\qquad\qquad\underset{3}{}\qquad\quad\underset{4}{}\qquad\underset{5}{}$$

The terms on the right-hand side of equation (5.38) represent the salt flux associated with:
   1. The mean river flow.
   2. The vertical 'gravitational' circulation.
   3. The horizontal circulation.
   4. Correlation of tidal fluctuations of velocity and salinity.
   5. Diffusion on the short period turbulence.

In the event of several stations on a cross-section being occupied simultaneously for a sufficient time, only the Term 5 will be unknown and K_x can be calculated. However, terms involving tidal fluctuation in cross-sectional areas are omitted in this analysis.

Bowden and Gilligan (1971) have regarded Term 3 as part of the turbulent diffusion term (T_D) with a suitably defined modified dispersion coefficient, and have assumed the tidal fluctuation cross-product (Term 4) as negligible. Thus equation (5.38) becomes:

$$0 = R\bar{s}_A + b \int_0^h \overline{u_1 s_1}\, dz - AK_x' \frac{\partial \bar{s}_A}{\partial x} \qquad (5.39)$$

or

$$0 = T_R - T_A - T_D$$

This expresses the downstream salt transport by the river discharge T_R as balanced by T_A, the advective upstream transport by the density current, the net vertical circulation, and by upstream diffusion, which includes the effect of the net transverse circulation. Consequently the diffusive fraction of the total upstream salt flux v (see chapter 8) is given by

$$v = \frac{T_D}{T_A + T_D} = \frac{T_R - T_A}{T_R}$$

Both T_R and T_A were computed directly for data from the Mersey. The calculated diffusive fractions are shown in Table 8.1. The value of v was least in the centre part of the Narrows, where the density current was most highly developed. The fraction increased towards either end of the Narrows where the estuary widens and becomes shallower. At the Liverpool Bay end, where the exchange of salt is probably carried out largely by horizontal eddies, the diffusive fraction was 0·85.

It is interesting to note that in this analysis the upstream salt flux by the density current (T_A) is advective, whereas in the previous method (equations 5.32 and 5.33) this term would be included in the diffusive part. This illustrates how the values obtained for the longitudinal dispersion coefficient depend on the method of analysis of the data.

In equation (5.38) the effect of cross-sectional deviation of the tidal fluctuations of velocity and salinity from the cross-sectional average is not considered. Thus Term 4 could be separated into two parts, one expressing the transverse and the other the vertical variation in the tidal fluctuations. This is done in the analysis of Fischer (1972). However, in this analysis the effect of fluctuations in cross-sectional area is not considered.

The observed velocity

$$u_o = u_A + u_d$$

The cross-sectional mean velocity undergoes a tidal fluctuation and can be represented by $u_A = \bar{u}_A + U_A$. Similarly $u_d = \bar{u}_d + U_d$. Considering a three-dimensional profile both $\bar{u}_d$ and U_d can be separated into variations in the vertical and transverse directions. Thus $\bar{u}_d = \bar{u}_{dt} + \bar{u}_{dv}$ and $U_d = U_t + U_v$. Salinity can be treated similarly. The mean salt flux over a tidal cycle through a unit area of the cross-section then is:

$$\overline{u_o s_o} = \bar{u}_A \bar{s}_A + \overline{U_A S_A} + (\bar{u}_{dt} \bar{s}_{dt})_A + (\bar{u}_{dv} \bar{s}_{dv})_A + \overline{(U_t S_t)_A} + \overline{(U_v S_v)_A} \quad (5.40)$$

In equation (5.40) the terms on the right-hand side represent the salt flux due to:

1. Mean flow on the river discharge.
2. Correlation of tidal fluctuations of sectional mean salinity and velocity.
3. Net transverse circulation.
4. Net vertical circulation.
5. Transverse oscillatory shear.
6. Vertical oscillatory shear.

Fisher (1972) has evaluated the contribution of the last four terms to the dispersion coefficient, from data for the Mersey. He concludes that the net transverse circulation is the dominant one, with the net vertical circulation and the vertical oscillatory terms an order of magnitude smaller. These results are additional to the true turbulent diffusion on the short period fluctuations and the net vertical circulation appears to be less important than the analysis of Bowden and Gilligan (1971) suggests. They calculated that about half of the longitudinal transport of salt was due to the net vertical circulation.

It is apparent that the effect of lateral and vertical differences in velocity and salinity can produce most of the upstream salt flux that is required to balance that carried downstream on the river flow. This effect is not purely diffusive in that it is not entirely caused by short period turbulent mixing, but it is due to differences in longitudinal advection coupled with vertical and lateral diffusion. A certain amount of the salt balance is also caused by the estuarine tidal response having a progressive element, so that terms produced by covariance with fluctuations of cross-sectional area have an appreciable magnitude. So far it is not possible to define precisely which are the dominant factors since different investigators have used slightly different methods of analysis; they split up their components in a variety of ways with certain implicit assumptions. Consequently, the results of differences in tidal response and topography between estuaries are not clear. A combination of the approaches of Hansen (1965) and Fischer (1972) would seem to be rewarding if applied to a number of well mixed estuaries. This would lead to equation (5.41).

Using the definitions

$$A = \bar{A} + A + A', \ u_A = \bar{u}_A + U_A + u'_A, \ s_A = \bar{s}_A + S_A + s'_A, \ u_d = \bar{u}_d + U_d + u'_d,$$

$$s_d = \bar{s}_d + S_d + s'_d \text{ where } \bar{u}_d = \bar{u}_{dt} + \bar{u}_{dv}, \ \bar{s}_d = \bar{s}_{dt} + \bar{s}_{dv} \text{ and } U_d = U_t + U_v$$

$$S_d = S_t + S_v \text{ and } u'_d = u'_{dt} + u'_{dv}, \ s'_d = u'_{dt} + u'_{dv}$$

Then the mean flux of salt through a section over a tidal cycle (equation 5.36) becomes, neglecting some terms:

$$\begin{aligned}
\bar{F}_s = {} & \bar{A}\bar{u}_A\bar{s}_A + \overline{AU_A}s_A + \overline{AU_A}S_A + \overline{AU_A}S_A + \bar{A}(\bar{u}_{dt}\bar{s}_{dt})_A + \bar{A}(\bar{u}_{dv}\bar{s}_{dv})_A \\
& + \overline{A(U_tS_t)}_A + \overline{A(U_vS_v)}_A + \overline{A(U_tS_t)}_A + \overline{A(U_vS_v)}_A + \overline{A(u'_{dt}s'_{dt})}_A + \\
& \overline{A(u'_{dv}s'_{dv})}_A
\end{aligned} \tag{5.41}$$

Of the terms on the right-hand side, Hansen (1965) found that Term 3 and Terms 5 and 6 were important, and Fischer (1972) found Term 5 the most important. Though Terms 11 and 12 are those relating to eddy–diffusion, all terms except the first would be included in the calculation of a longitudinal dispersion coefficient with a formula similar to equation (5.33).

Considerable use has been made of observations on the spreading of dye patches in order to calculate the dispersion caused by turbulence in the sea and in estuaries. In particular Rhodamine B has proved the most favourable (Pritchard and Carpenter, 1960), being detectable at low concentrations by towed fluorometers. In most cases instantaneous rather than continuous release is used. In a uniform flow field the patch of dye spreads so that after an initial period the distribution becomes Gaussian in form and, because of dispersion, the patch becomes larger with time. The variance of the dye distribution σ_{rc}^2 is proportional to t^2-t^3. An apparent diffusivity K_a can be defined as $K_a = \sigma_{rc}^2/4t$ and this varies with the scale of the diffusion l defined as $3\sigma_{rc}$. Results show that K_a is proportional to $l^{\frac{4}{3}}$ over ranges of l from 10 m to 10000 km (Okubo, 1971). In stratified water and where there is velocity shear the relationships are not clear though the mixing coefficients decrease with increasing stratification and Richardson number.

Because the turbulence terms cannot be independently measured at the moment, but can only be estimated as the difference of summation of the other terms, solution of general salt balance equation is impossible. Consequently, simplifying assumptions have to be made and the most suspect one is that of lateral homogeneity. Solution may eventually be possible if instrumentation and finance allow the direct measurement of the various turbulent fluxes, in addition to the advective ones. The measurement of the spectrum of the turbulent terms, such as $u's'$, is of great priority in order that the necessary averaging processes can be carried out properly.

Once these problems are solved, it will be necessary to use the same basic methods of gathering and analysing data on a number of estuaries before any more general conclusions can be stated. At present, bearing in mind the qualifications introduced by lateral averaging, it appears that Pritchard's (1955) analysis of the importance of the various terms in the different estuarine types is broadly correct. However, the consideration of the shear effect by Bowden (1963), Hansen (1965) and Fischer (1972) appears to be a very useful and realistic approach to the problem as it emphasizes the contribution of gravitational and transverse circulation, an effect that is not apparent in the other analyses. In particular it is leading up to a better understanding of the processes causing pollutant dispersion in estuaries, and the dispersion coefficients obtained enable a more realistic prediction of flushing and dispersal of effluents to be made.

CHAPTER 6

Dynamic Balance

GENERAL FORMULATION

In the last chapter we examined the salt balance in an estuary and the water flows required to produce this balance. There are other constraints, however, because the water flows, as well as satisfying the salt balance, must also be acting according to a dynamic balance. The density distribution produces pressure gradients and these are a major force controlling the current flows. Turbulent eddies produced by the flows cause exchanges of momentum, as well as of salt, and produce frictional forces helping to resist the flow.

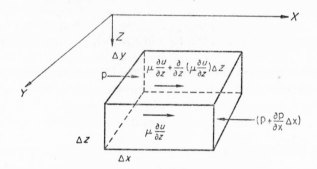

Figure 6.1   Forces acting on an element

The dynamic balance is provided by Newton's second law of motion, which states that force equals mass times acceleration. The forces involved should include: the sea surface gradients and internal density gradients which give horizontal differences in pressure, Coriolis force due to the earth's rotation, wind stresses on the sea surface and frictional forces on the sea bed, turbulent stresses or eddy viscosity and molecular viscosity. The accelerations will be comprised of terms of the form $\partial u/\partial t$ and $u\partial u/\partial x$.

Consider an element with sides of lengths Δx, Δy and Δz (Figure 6.1) in a co-ordinate system where the x axis is positive down the estuary, the y axis is directed across the estuary, positive to the right, and the z axis is vertically downwards. The forces acting on the element are composed of surface forces and body forces. The surface forces consist of two parts, one acting parallel to the surfaces of the element (shear) and the other acting normal to the surface (pressure). Body forces depend on the mass of fluid in the element and are caused by gravity and the Coriolis effect.

The shearing stress is the means by which the water above a horizontal plane acts on that below. The frictional forces are derived from the vertical changes in the horizontal shearing stress. If the stress is constant with depth there is no frictional force and if the stress varies linearly with depth, then the frictional force is constant.

The frictional force on the bottom of the element in the x direction at any instant is

$$\mu \frac{\partial u}{\partial z} \Delta x \Delta y$$

where μ is the coefficient of molecular viscosity. On the top of the element, by Taylors' series, the frictional force is

$$\mu \frac{\partial u}{\partial z} \Delta x \Delta y + \frac{\partial}{\partial z}\left(\mu \frac{\partial u}{\partial z}\right)\Delta z \Delta x \Delta y$$

This component of the total frictional force is the difference:

$$\frac{\partial}{\partial z}\left(\mu \frac{\partial u}{\partial z}\right)\Delta z \Delta x \Delta y = \mu \frac{\partial^2 u}{\partial z^2}\Delta z \Delta x \Delta y$$

Similarly the component of the total frictional force at any instant caused by friction on the sides of the element will be

$$\mu \frac{\partial^2 u}{\partial y^2}\Delta y \Delta x \Delta z$$

and on the ends of the element

$$\mu \frac{\partial^2 u}{\partial x^2}\Delta x \Delta y \Delta z$$

The net pressure force on the element in the x direction is

$$p\Delta y \Delta z - \left(p + \frac{\partial p}{\partial x}\Delta x\right)\Delta y \Delta z = -\frac{\partial p}{\partial x}\Delta x \Delta y \Delta z$$

In the x direction the Coriolis force is $(2\omega \sin \phi v - 2\omega \cos \phi \cos \theta w)\rho\Delta x \Delta y \Delta z$ where θ is the angle between the positive x axis and east, ϕ is the latitude and ω the angular velocity of the earth's rotation. The Coriolis force is an apparent force that is introduced into the equation of motion to allow a frame of reference to be used that is fixed relative to the rotating earth. The component due to the vertical velocity is maximum when the x axis lies along the east–west direction, and zero when directed along a north–south line.

Now, Newton's second law of motion gives us: Mass × acceleration = surface forces + body forces.

Thus:

$$\rho \Delta x \Delta y \Delta z \left(\frac{\partial u}{\partial t} + u \frac{\partial u}{\partial x} + v \frac{\partial u}{\partial y} + w \frac{\partial u}{\partial z} \right) =$$

$$- \frac{\partial p}{\partial x} \Delta x \Delta y \Delta z + (2\omega \sin \phi v - 2\omega \cos \phi \cos \theta w) \rho \Delta x \Delta y \Delta z$$

$$+ \mu \Delta x \Delta y \Delta z \left(\frac{\partial^2 u}{\partial x^2} + \frac{\partial^2 u}{\partial y^2} + \frac{\partial^2 u}{\partial z^2} \right)$$

Therefore the instantaneous longitudinal equation of motion is:

$$\frac{\partial u}{\partial t} + u \frac{\partial u}{\partial x} + v \frac{\partial u}{\partial y} + w \frac{\partial u}{\partial z} = - \frac{1}{\rho} \frac{\partial p}{\partial x} + f_1 v - f_2 w \cos \theta$$

$$+ \frac{\mu}{\rho} \left(\frac{\partial^2 u}{\partial x^2} + \frac{\partial^2 u}{\partial y^2} + \frac{\partial^2 u}{\partial z^2} \right) \tag{6.1}$$

Similarly the vertical dynamic balance is:

$$\frac{\partial w}{\partial t} + u \frac{\partial w}{\partial x} + v \frac{\partial w}{\partial y} + w \frac{\partial w}{\partial z} = - \frac{1}{\rho} \frac{\partial p}{\partial z} + g - f_2(u \cos \theta - v \sin \theta)$$

$$+ \frac{\mu}{\rho} \left(\frac{\partial^2 w}{\partial x^2} + \frac{\partial^2 w}{\partial y^2} + \frac{\partial^2 w}{\partial z^2} \right) \tag{6.2}$$

Where $(u \cos \theta - v \sin \theta)$ is the east–west component of the horizontal flows and g the gravitational acceleration.

The lateral balance is:

$$\frac{\partial v}{\partial t} + u \frac{\partial v}{\partial x} + v \frac{\partial v}{\partial y} + w \frac{\partial v}{\partial z} = - \frac{1}{\rho} \frac{\partial p}{\partial y} - f_1 u - f_2 w \sin \theta$$

$$+ \frac{\mu}{\rho} \left(\frac{\partial^2 v}{\partial x^2} + \frac{\partial^2 v}{\partial y^2} + \frac{\partial^2 v}{\partial z^2} \right) \tag{6.3}$$

VERTICAL DYNAMIC BALANCE

In equation (6.2) the vertical Coriolis effect is negligible compared with the gravitational acceleration and, if turbulent fluctuations are averaged out by taking time means, the turbulent stresses and the vertical acceleration terms are normally assumed negligible also. Thus equation (6.2) becomes the hydrostatic equation

$$\overline{\frac{1}{\rho} \frac{\partial p}{\partial z}} = g \tag{6.4}$$

The hydrostatic assumption is universally made both in the open ocean and in estuaries.

LONGITUDINAL DYNAMIC BALANCE

The horizontal, lateral and vertical velocities can each be considered as the sum of three constituents; a time mean over a tidal cycle, a simple harmonic velocity fluctuation of tidal period and a short period turbulent fluctuation. Substitution in equation (6.1) gives:

$$\frac{\partial \bar{u}}{\partial t} + \bar{u}\frac{\partial \bar{u}}{\partial x} + \bar{v}\frac{\partial \bar{u}}{\partial y} + \bar{w}\frac{\partial \bar{u}}{\partial z} + \frac{\partial}{\partial x}(\overline{UU}) + \frac{\partial}{\partial y}(\overline{UV}) + \frac{\partial}{\partial z}(\overline{UW})$$

$$= -\left(\frac{1}{\rho}\frac{\partial p}{\partial x}\right) + f_1 v - \frac{\partial}{\partial x}\overline{(u'u')} - \frac{\partial}{\partial y}\overline{(u'v')} - \frac{\partial}{\partial z}\overline{(u'w')} \qquad (6.5)$$

The molecular stress terms will be negligible compared with the turbulent ones. Because the vertical velocities are small the second Coriolis term can also be neglected. The various other cross-products have been neglected for reasons similar to those considered for the salt balance.

In equation (6.5) the non-linear inertial terms associated with the tidal motion, such as $\partial/\partial x(\overline{UU})$, arise because the mean value of U^2 over a tide is not zero. If U is a simple harmonic function then $\overline{U^2} = U_o^2 \cos^2 \omega t = \frac{1}{2}U_o^2$ and $\partial/\partial x(\overline{UU}) = U_o \partial U_o/\partial x$. If the tidal wave is a purely progressive one then U and W will be 90° out of phase and the term $\partial/\partial z(\overline{UW})$ disappears. This term could also be neglected if the amplitudes of the tidal motions do not vary significantly with depth. This is unlikely, however, as the amplitude of both U and W will be reduced near the bottom. In a standing wave system the vertical and horizontal tidal velocities will be in phase and consequently $\overline{UW} = \frac{1}{2}U_o W_o$. Then

$$\frac{\partial}{\partial x}(\overline{UU}) + \frac{\partial}{\partial z}(\overline{UW}) = \frac{1}{2}\left(U_o\frac{\partial U_o}{\partial x} - W_o\frac{\partial U_o}{\partial z}\right)$$

The term $\partial/\partial y(\overline{UV})$ disappears if lateral homogeneity is assumed.

The pressure term $\overline{((1/\rho)\partial p/\partial x)}$ is normally taken to be represented by $((1/\bar{\rho})\partial\bar{p}/\partial x)$ which assumes that pressure and density fluctuations in the water are uncorrelated. The pressure at any depth is composed of two parts, one part due to the density distribution and the other due to the slope of the free surface. The density distribution is obtained from salinity and temperature measurements but the surface slope is normally obtained by calculation using a known or assumed level at which $\partial\bar{p}/\partial x$ is zero.

From the hydrostatic equation (6.4), the pressure at any depth z is

$$p = p_a + g\int_{-\zeta}^{z} \rho \, dz$$

where p_a is the atmospheric pressure, which can be assumed uniform and ζ is the surface elevation.

Thus the horizontal pressure gradient at any depth z is

$$\frac{\partial p}{\partial x} = g \int_{-\zeta}^{z} \frac{\partial \rho}{\partial x} \, dz - g\rho_s \frac{\partial \zeta}{\partial x} \tag{6.6}$$

where ρ_s is the density at the sea surface.

In a homogeneous sea the first term on the right of equation (6.6) is zero and the pressure gradient is caused by the sea surface slope. As a consequence the surfaces of equal pressure are non-level throughout the depth. In stratified conditions the integral term increases with depth so that the contribution of the sea surface slope is gradually compensated. Flows induced by the first, density, part of equation (6.6) are called baroclinic flows and those due to the sea surface gradient are barotropic flows.

The integral term on the right-hand side of equation (6.6) can be written in the form normally used in dynamical oceanography by replacing the density by specific volume $\alpha = 1/\rho$ and integrating with respect to p.

Thus

$$\int_{-\zeta}^{z} \frac{\partial \rho}{\partial x} \, dz = -\frac{\langle \rho \rangle}{g} \frac{\partial}{\partial x} \int_{p_a}^{p_z} \alpha \, dp$$

Where $\langle \rho \rangle$ is the mean density between the surface and the depth z, and equation (6.6) becomes:

$$\frac{1}{\rho} \frac{\partial p}{\partial x} = -\frac{\partial}{\partial x} \int_{p_a}^{p_z} \alpha \, dp - g\frac{\partial \zeta}{\partial x} \tag{6.7}$$

The integral on the right-hand side of equation (6.7) is the expression for the dynamic depth of the pressure p_z

$$\mathbf{D} = \int_{p_a}^{p_z} \alpha \, dp$$

The second term, which is a constant for any particular situation, can be calculated if $\partial p/\partial x = 0$ at a depth $z = H$. Then from equation (6.6)

$$-g\frac{\partial \zeta}{\partial x} = -\frac{g}{\rho_s} \int_{-\zeta}^{H} \frac{\partial \rho}{\partial x} \, dz = \frac{\langle \rho \rangle}{\rho_s} \frac{\partial}{\partial x} \int_{p_a}^{p_H} \alpha \, dp$$

Substituting in equation (6.7) gives:

$$\frac{1}{\rho} \frac{\partial p}{\partial x} = -\frac{\partial \mathbf{D}}{\partial x} + \text{constant} \quad \text{and} \quad \frac{1}{\bar{\rho}} \frac{\partial \bar{p}}{\partial x} = -\frac{\partial \bar{\mathbf{D}}}{\partial x} + C_1 \tag{6.8}$$

The horizontal gradient of the dynamic depth $\bar{\mathbf{D}}$ can be calculated from the distribution of tidal mean salinity and temperature with depth using standard oceanographic tables, such as La Fond (1960).

The final three terms on the right-hand side of equation (6.5) are the frictional forces due to turbulent effects. In order to get them in a form analogous to that for molecular viscosity they are written in the form:

$$\bar{\tau}_{xx} = \rho(\overline{u'u'}) = -\rho N_x \frac{\partial \bar{u}}{\partial x}$$

$$\bar{\tau}_{xz} = \rho(\overline{u'w'}) = -\rho N_z \frac{\partial \bar{u}}{\partial z}$$

$$\bar{\tau}_{xy} = \rho(\overline{u'v'}) = -\rho N_y \frac{\partial \bar{u}}{\partial y} \tag{6.9}$$

These turbulent stresses are called the Reynolds stresses and they are so much larger than the molecular ones that the latter can be neglected. The coefficients N_x, N_z and N_y are the coefficients of eddy viscosity in the respective directions. They have the same dimensions (cm^2 sec^{-1}) as the effective eddy–diffusion coefficients and as they involve tidal average values they should be termed effective coefficients of eddy viscosity. It is generally possible to neglect the term involving N_x. Also, by assuming lateral homogeneity, so that $\partial \bar{u}/\partial y$ is zero, the turbulent frictional terms in equation (6.5) reduce to

$$-\frac{\partial}{\partial z}(\overline{u'w'}) = -\frac{1}{\rho}\frac{\partial \bar{\tau}_{xz}}{\partial z} = \frac{\partial}{\partial z}\left(N_z \frac{\partial \bar{u}}{\partial z}\right) \tag{6.10}$$

At the surface and bottom the values of $\bar{\tau}_{xz}$ will be given by the longitudinal component of the net flux of momentum across the boundaries. At the surface it will be equal to the component of the mean wind stress and at the bottom equal to the mean bed shear stress ($\bar{\tau}_o$). It is common in a neutrally stratified tidal situation, to write the bed stress shear in terms of a frictional coefficient k times the square of the depth mean velocity. Thus $\tau_o = k\langle u_o\rangle|\langle u_o\rangle|$. The modulus form $|\langle u_o\rangle|$ is required to preserve the sign of the oscillating current. Values of k are normally in the range $0\cdot002$–$0\cdot005$.

In a current of uniform density the shearing stress in a steady flow will be linear from surface to bottom, but with a time varying flow no direct inference can be drawn about the distribution of shearing stress with depth. However, Bowden *et al.* (1959) have examined the distribution of shearing stress in a tidal flow with uniform density in Red Wharf Bay. In this context the equation of motion for a depth z becomes:

$$\frac{\partial u_o}{\partial t} = -g\frac{\partial \zeta}{\partial x} + \frac{1}{\rho}\frac{\partial \tau_{xz}}{\partial z} \tag{6.11}$$

Taking depth mean values and assuming that the wind stress is zero,

$$\frac{\partial \langle u\rangle}{\partial t} = -g\frac{\partial \zeta}{\partial x} - \frac{\tau_o}{\rho h} \tag{6.12}$$

The difference between equations (6.11) and (6.12) gives

$$\frac{\partial}{\partial t}(u_o - \langle u \rangle) = \frac{1}{\rho}\frac{\partial \tau_{xz}}{\partial z} + \frac{\tau_o}{\rho h}$$

For finite time increments Δ_t

$$\frac{1}{\rho}\frac{\partial \tau_{xz}}{\partial z} = -\frac{\tau_o}{\rho h} + A_x \qquad \text{where } A_x = \frac{\Delta_t(u_o - \langle u \rangle)}{\Delta_t} \qquad (6.13)$$

Integrating (6.13) with respect to z

$$\tau_{xz} = -\tau_o\frac{z}{h} + \rho \int_0^z A_x \, dz \qquad (6.14)$$

Thus the distribution of τ_{xz} with depth can be calculated from a knowledge of the bottom stress and the acceleration term A_x. The bottom stress can be calculated from the velocity profile using the von-Karman–Prandtl formula:

$$u_o = \frac{1}{k_o}\left(\frac{\tau_o}{\rho}\right)^{\frac{1}{2}} \ln\frac{z}{z_o} \qquad (6.15)$$

where z_o is the bottom roughness length (the height above the bed at which the velocity is zero) and k_o is the von-Karman constant $= 0\cdot4$. The velocity is measured at a height z above the bottom. Though derivation of this formula assumes a constant shear stress above the bottom, and neutral conditions, profiles fitting the formula are commonly observed within 2 m or so of the bed.

In Red Wharf Bay, Bowden, *et al.* (1959) have applied the above treatment to half-hourly data. They found the bed shear stress varied cyclicly with maximum values of about 8 dynes cm^{-2}. The curves of τ_{xz} as a function of depth showed an almost linear increase from surface to bottom at times near maximum flood and ebb when the acceleration terms were small. When the flow was accelerating and the stress increasing, the stresses at mid-depths were less than those corresponding to a linear variation, and the curve of stress against depth was concave upwards. When the current was decelerating this effect was reversed, the stress at intermediate depths was greater than a linear variation and the curves were convex upwards.

The values of the coefficients of eddy viscosity were also calculated. The highest values occurred at mid-depth when the current was greatest, with a maximum value of about 500 cm^2 sec^{-1}. On the flood tide the mean value of N_z was about 270 cm^2 sec^{-1} and on the ebb about 130 cm^2 sec^{-1}. On dimensional grounds the maximum value of N_z would be proportional to the tidal velocity times the depth. In this case the average value at mid depth of $N_z = 2\cdot5 \times 10^{-3} Uh$. These short term eddy coefficients cannot be directly compared with those obtained from results averaged over a tidal period, which have a slightly different physical meaning.

In neutral conditions $K_z \approx N_z$, but when a density stratification is present the vertical turbulence is inhibited and both K_z and N_z will be reduced. As the rate of generation of turbulent energy must exceed the rate of increase of potential energy due to vertical mixing K_z will be reduced more than N_z.

Silver Bay

McAlister *et al.* (1959) have analysed the longitudinal dynamic balance of a typical Alaskan fjord. The non-linear tidal inertial term $U_o \partial U_o / \partial x$ was calculated from the volume of intertidal water entering the Bay. U_o was found to be $1 \cdot 9$ cm sec^{-1} at section 2, the small values being due to the large depth. This value of the tidal amplitude was an order smaller than the surface mean flow. The change in U_o along the channel was also small as the decrease in cross-sectional area compensated for the change in intertidal volume along the channel. Consequently this term was small enough to be neglected. The term $\partial / \partial z (\overline{UW})$ was not considered.

The horizontal component of the turbulent flux of momentum was considered negligible. Lateral homogeneity and zero lateral velocities were assumed.

The longitudinal dynamic balance thus became:

$$\bar{u}\frac{\partial \bar{u}}{\partial x} + \bar{w}\frac{\partial \bar{u}}{\partial z} = -\frac{1}{\rho}\frac{\partial p}{\partial x} - \frac{\partial}{\partial z}(\overline{u'w'}) \qquad (6.16)$$

The term $\bar{u}(\partial \bar{u}/\partial x)$ was calculated from the velocities derived using an advective salt balance and a depth of no motion at $5 \cdot 15$ m depth, as discussed in chapter 5. The term $\bar{w}(\partial \bar{u}/\partial z)$ was calculated by applying continuity principles to the calculated horizontal velocities. The pressure term was calculated assuming a depth of no motion at 100 m, a depth at which the mean horizontal velocity was zero. The water slope was $-1 \cdot 2 \times 10^{-6}$ in July and $-2 \cdot 1 \times 10^{-7}$ in March. As the surface wind stress was not known, $(\overline{u'w'})$ was taken as zero where $\partial \bar{u}/\partial z = 0$, at 100 m and at 12 m depth in July and 18 m depth in March. All terms in equation (6.16) can then be calculated.

The values derived for the terms under high and low discharge conditions are shown in Table 6.1. It appears that below about 10 m depth the terms are virtually immeasurably small. The coefficient of vertical eddy viscosity N_z was calculated for these results. The values were 3–30 cm^2 sec^{-1} for the high river flow period (July) and 5–120 cm^2 sec^{-1} for the low flow period (March).

The July, high river discharge, circulation was largely driven by the addition of fresh water into the inlet, while the March circulation appears to have been caused largely by the replacement of water at depth in Silver Bay. In both summer and winter there was a strong pressure gradient associated with the surface slope, with the largest gradient limited to the upper 5 m. In winter the level pressure surface was associated with the level of no net motion at about 9 m depth. Below this the balance was between the pressure gradient and the vertical stress gradient. In summer the pressure field did not reverse and even

TABLE 6.1

Longitudinal Dynamic Balance in Silver Bay [a]

z	$1/\bar{\rho}(\partial\bar{p}/\partial x)$		$\bar{u}(\partial\bar{u}/\partial x)$		$\bar{w}(\partial\bar{u}/\partial z)$		$\partial/\partial z\overline{(u'w')}$	
			cm sec$^{-2}\times 10^5$					
(m)	1	2	1	2	1	2	1	2
0	−120	−20	150	18	0	0	−30	2
1	−70	−5	90	9	35	3	−55	−7
2	−25	0	37	4	52	5	−64	−9
3	−11	2	17	1	45	4	−51	−7
4	−6	4	4	0	34	4	−32	−8
5	−6	4	0	0	27	2	−21	−6
6	−5	3	0	0	22	0	−17	−3
7	−2	2	2	0	19		−17	−2
8	−1	1	4	0	15		−18	−1
9	−0·5	0	7	0	9		−15	0
10	−0·5	−1	11	0	0		−11	1
15	−0·5	−3	5	0	−1		−4	3
20	−0·5	−4	2	0	0		−2	4
25	−0·5	−4	1	0	0		−1	4

1. High flow—July   2. Low flow—March

[a] Reproduced with permission from W. B. McAlister, M. Rattray and C. A. Barnes, 1959, *Tech. Rept.* 62, University of Washington, Dept. Oceanography, Tables 4 and 5.

in the inflowing layer the pressure appeared to be greater inside Silver Bay than outside. The inertial terms were important in the upper, faster moving, layers.

James River

Pritchard (1956) has examined the dynamic balance in the James River. He has considered steady-state conditions and zero lateral velocities. It was argued from the salt balance analysis that the horizontal eddy flux of salt was small and Pritchard assumes by analogy that the horizontal eddy flux of momentum was also negligible. Consequently equation (6.5) becomes:

$$\bar{u}\frac{\partial\bar{u}}{\partial x}+\bar{w}\frac{\partial\bar{u}}{\partial z}+U_0\frac{\partial U_0}{\partial x} = -\frac{\partial\bar{D}}{\partial x}+C_1-\frac{\partial}{\partial z}\overline{(u'w')} \qquad (6.17)$$

From the velocity data the acceleration terms on the left-hand side were calculated and from the salinity and temperature distribution the relative pressure forces also. To solve the equation for $\bar{\tau}_{xz}$, two boundary conditions are required to resolve the constant C and the constant of integration of the Reynolds stress term.

At the surface and the bottom $\bar{\tau}_{xz}$ will be equal to the net flux of momentum across these boundaries. At the surface, if the mean wind stress is zero then the momentum flux will be zero. At the bottom $\bar{\tau}_{xz} = \bar{\tau}_o$, where τ_o is the shear stress on the bed. If the sea bed is hydrodynamically rough the von Karman–Prandtl equation can be used:

$$\tau_o = \rho\left(\frac{u_o k_o}{\ln z/z_o}\right)^2 \qquad (6.18)$$

where k_o, von Karman's constant, equals 0·4, and the bottom roughness z_o is assumed. For the mean bed shear stress the mean velocity at a height z above the bottom should not be used. A good approximation should be obtained using the root mean square of the peak ebb and flood velocities, or, if the tidal variation can be considered as a trigonometrical function, then $\overline{u_z^2} = \frac{1}{2}U_o^2$. Rather large variations in $\bar{\tau}_o$ and consequently in the calculated $\bar{\tau}_{xz}$, and the sea surface slope, can be obtained for fairly small variations in z_o. Pritchard assumed a roughness length of 0·02 cm.

Evaluation of the constant C_1 enables determination of the depth of the level pressure surface and the absolute pressure field distribution. In the James River the level pressure surface existed near mid-depth. Above this depth the pressure surfaces sloped towards the sea and below it they sloped towards the head of the estuary. The longitudinal level pressure surfaces

TABLE 6.2

Magnitudes of the Terms in the Longitudinal Dynamic Balance of the James River [a]

Depth (m)	$\bar{u}(\partial\bar{u}/\partial x)$	$\bar{w}(\partial\bar{u}/\partial z)$	$U_o(\partial U_o/\partial x)$	$-\overline{\alpha(\partial p/\partial x)}$	$\partial/\partial z\overline{(u'w')}$
			m sec$^{-2} \times 10^6$		
0	0·03	0·01	2·97	14·30	11·29
1	0·12	0·09	2·97	10·56	7·38
2	0·08	0·28	2·97	6·83	3·50
3	0·01	0·42	2·97	3·10	−0·30
4	0·09	0·20	2·97	−0·64	−3·90
5	0·11	0·04	2·97	−4·38	−7·50
6	−0·04	0·01	2·97	−8·09	−11·03
7	−0·04	0·02	2·97	−11·77	−14·72

[a] Reproduced with permission from D. W. Pritchard, *Jour. Marine Res.*, **15**, 1956, Table 2.

were above or below the depth of no mean motion depending mainly on the sign and the magnitude of the term $U_o\partial U_o/\partial x$. When this term was small the level pressure surface was nearly the same as the depth of no mean motion. If the tidal velocity amplitude U_o increased downstream then the depth would be greater than the depth of the no mean motion and vice versa.

As the Reynold's stress term is obtained by summing the other terms in equation (6.17) it includes accumulated errors, and the effects of terms not otherwise considered. The magnitudes of the various terms in equation (6.17) for the James River are shown in Table 6.2. The longitudinal component of the pressure force was mainly balanced by the eddy frictional term. However, a significant portion of the pressure force was balanced by the field change in the amplitude of the tidal velocity. This contrasts with the situation in fjords. The various longitudinal acceleration terms appear to be insignificant, also contrasting with the results in the surface layer of fjords.

Mersey Narrows

An approach similar to that used in Red Wharf Bay (p. 95) has been used by Bowden (1960) and Bowden and Sharaf el Din (1966a) in the Mersey Narrows. Assuming that lateral velocities are small and that $f_1 v$ is negligible, the equation of motion in the longitudinal sense at any depth can be written as:

$$\frac{\partial u_o}{\partial t} = -g\frac{\partial \zeta}{\partial x} - gP - \frac{1}{\rho}\frac{\partial \tau_{xz}}{\partial z} \tag{6.19}$$

where

$$P = \frac{1}{\rho} \int_0^z \frac{\partial \rho}{\partial x}.\,dz$$

Integrating from $z = 0$ to $z = h$ and assuming the surface stress to be zero

$$\frac{\partial \langle u \rangle}{\partial t} = -g\frac{\partial \zeta}{\partial x} - g\langle P \rangle - \frac{\tau_o}{\rho h} \tag{6.20}$$

Taking differences between equations (6.19) and (6.20)

$$\frac{\partial(u_o - \langle u \rangle)}{\partial t} = -g(P - \langle P \rangle) - \frac{1}{\rho}\frac{\partial \tau_{xz}}{\partial z} + \frac{\tau_o}{\rho h}$$

and   (6.21)

$$\frac{\partial(u_o - \langle u \rangle)}{\partial t} = \frac{\partial u_1}{\partial t} \quad\text{as}\quad u_o = \langle u \rangle + u_1$$

An equation similar to (6.21) can be developed which applies to the mean values over a tidal cycle. In the steady state if $\partial \bar{u}_1/\partial t = 0$, then:

$$\bar{\tau}_{xz} = -g \int_0^z \rho(\bar{P} - \langle \bar{P} \rangle)\,dz + \frac{\bar{\tau}_o z}{h} = -\rho N_z \frac{\partial \bar{u}_1}{\partial z} \tag{6.22}$$

Bowden (1960) has applied this approach to data obtained at a station in the centre of the Mersey Narrows during two periods. The values of $(\bar{P} - \langle \bar{P} \rangle)$ were determined from the density distribution and the shearing stress $\bar{\tau}_{xz}$ on the first occasion was taken as zero when $\partial \bar{u}_1/\partial z = 0$. This occurred at a depth z/h of 0·75. On the second occasion $\bar{\tau}_{xz}$ was taken as zero at the bottom. The values calculated for $\bar{\tau}_{xz}$ and N_z given in Table 6.3 are values related to the tidal cycle means and could be termed the effective shear stress and effective vertical eddy viscosity. Maximum values occur at mid-depth, but are only about 1/10 of those which one would expect in a tidal current of uniform density. Values of the effective eddy–diffusion coefficient K_z were

TABLE 6.3

Values of Effective Shearing Stress, Eddy-Viscosity and Eddy–Diffusion Coefficients for Two Periods in the Mersey Narrows [a]

	1st period			2nd period		
z/h	$\bar{\tau}_{xz}$ dynes cm^{-2}	N_z cm^2 sec^{-1}	K_z cm^2 sec^{-1}	$\bar{\tau}_{xz}$ dynes cm^{-2}	N_z cm^2 sec^{-1}	K_z cm^2 sec^{-1}
0·1	0·2	9	5	0·34	14	8
0·3	0·45	27	11	0·86	46	23
0·5	0·39	40	27	1·03	73	30
0·7	0·10	43	17	0·86	72	29
0·9	−0·42	62	3	0·34	25	13

[a] Reproduced with permission from K. F. Bowden, *I.A.S.H. Comm. Surface Waters*, Publ. 51, 1960, Table 3.

calculated for the same data (Table 6.3) using equation (5.25), and are about half the corresponding N_z. During the tidal cycle, however, instantaneous values for N_z and for K_z are likely to be much larger than these values.

Further data from the Mersey has been treated in the same way by Bowden and Sharaf el Din (1966a). In this case the Coriolis term was retained and the value of $\bar{\tau}_o$ was estimated on the assumption that $\bar{\tau}_{xz}$ was zero when $\partial \bar{u}/\partial z = 0$. This occurred at a depth $z/h = 0·8$. The calculated values of the effective shearing stress and the effective coefficient of vertical eddy-viscosity, for stations 1–3 (Figure 4.18), are given in Table 6.4. The values of N_z tended to

TABLE 6.4

Effective $\bar{\tau}_{xz}$, N_z and K_z for the Mersey Narrows [a]

	Station 1			Station 2			Station 3			Mean across section	
z/h	$\bar{\tau}_{xz}$ dynes cm^{-2}	N_z cm^2 sec^{-1}	K_z	$\bar{\tau}_{xz}$ dynes cm^{-2}	N_z cm^2 sec^{-1}	K_z	$\bar{\tau}_{xz}$ dynes cm^{-2}	N_z cm^2 sec^{-1}	K_z	$\bar{\tau}_{xz}$ dynes cm^{-2}	N_z cm^2 sec^{-1}
0·1	0·03	1[b]	8	0·31	240[b]	6	0·19	8	5	0·14	8·5
0·3	0·07	15	12	0·69	4	15	0·42	14	13	0·33	20
0·5	0·11	8	22	0·60	23	28	0·37	14	17	0·35	27
0·7	0·09	23	22	0·28	113	23	0·16	11	9	0·19	29
0·9	−0·11	5	1	−0·30	49	3	−0·16	22	1	−0·21	20

[a] Reproduced with permission from K. F. Bowden and S. H. Sharaf el Din, *Geophy. Jour.*, **10**, 1966. Tables 3 and 4.
[b] Figures are uncertain and omitted from mean.

reach a maximum at mid-depth, but more consistent results were obtained by using cross-sectional mean values in equation (6.22). These values of N_z are somewhat greater than the corresponding values of the effective vertical eddy–diffusion coefficient calculated using equation (5.25).

LATERAL DYNAMIC BALANCE

For average values over a tidal cycle the lateral dynamic balance will be

$$\frac{\partial \bar{v}}{\partial z} + \bar{u}\frac{\partial \bar{v}}{\partial x} + \bar{v}\frac{\partial \bar{v}}{\partial y} + \bar{w}\frac{\partial \bar{v}}{\partial z} + \frac{\partial}{\partial x}(\overline{VU}) + \frac{\partial}{\partial y}(\overline{VV}) + \frac{\partial}{\partial z}(\overline{VW})$$

$$= -\frac{1}{\bar{\rho}}\frac{\partial \bar{p}}{\partial y} - f_1\bar{u} - f_2\bar{w}\sin\theta - \frac{\partial}{\partial x}(\overline{v'u'}) - \frac{\partial}{\partial y}(\overline{v'v'}) - \frac{\partial}{\partial z}(\overline{v'w'}) \qquad (6.23)$$

In any estuary the lateral and longitudinal tidal fluctuations are likely to be in phase so that in a purely progressive tidal wave situation the tidal terms will be

$$\frac{1}{2}\left(U_o\frac{\partial U_o}{\partial x} + V_o\frac{\partial V_o}{\partial y}\right)$$

For a standing wave they will be

$$\frac{1}{2}\left(U_o\frac{\partial U_o}{\partial x} + V_o\frac{\partial V_o}{\partial y} - W_o\frac{\partial V_o}{\partial z}\right)$$

However, these terms are generally not considered. The vertical Coriolis term is also considered negligible. The Reynolds stresses in this case are written:

$$\bar{\tau}_{yx} = \rho(\overline{v'u'}) = -\rho N_x\frac{\partial \bar{v}}{\partial x}$$

$$\bar{\tau}_{yy} = \rho(\overline{v'v'}) = -\rho N_y\frac{\partial \bar{v}}{\partial y}$$

$$\bar{\tau}_{yz} = \rho(\overline{v'w'}) = -\rho N_z\frac{\partial \bar{v}}{\partial z}$$

The dominant term of the three is likely to be that involving $\bar{\tau}_{yz}$. The pressure term can be again related to a depth of no motion, but it may not necessarily be the same as that in the horizontal sense. As the lateral sea surface slope is different from that in the longitudinal sense the constant will be different.

Consequently, for steady-state conditions

$$\bar{u}\frac{\partial \bar{v}}{\partial x} + \bar{v}\frac{\partial \bar{v}}{\partial y} + \bar{w}\frac{\partial \bar{v}}{\partial z} = -\frac{\partial \overline{D}}{\partial y} + C_2 - f_1\bar{u} - \frac{1}{\rho}\frac{\partial \bar{\tau}_{yz}}{\partial z} \qquad (6.24)$$

In the lateral case, however, the curvature of the streamlines of the flow along the estuary is likely to be significant. The curvature is the result of the meandering shape of the estuary and gives rise to centripedal accelerations that can be thought of as being produced by a centrifugal force acting normal to the streamlines. The centrifugal force will equal the velocity squared divided by the radius of curvature. In terms of the mean flow the centrifugal force will be

$$\frac{\bar{u}^2 + \overline{U^2}}{r_{xy}}$$

where r_{xy} is the radius of curvature of the streamlines of the longitudinal flow in the y direction.

Providing the coordinate axes are chosen correctly, so that the radius of curvature is not at an angle to the y axis, then the accelerations on the left-hand side of equation (6.24) should be numerically equal to

$$\frac{\bar{u}^2 + \overline{U^2}}{r_{xy}}$$

Similarly the terms on the right-hand side of equation (6.24) should also equal the centrifugal force.

In order to see whether curvature of streamlines consistent with the estuarine topography is important in the balance between the various forces, a thin near-surface layer can be considered. If the gradient of the shearing stresses is constant with depth, then the terms involving the sea surface slope and the Reynolds stresses can be eliminated by considering the differences between each term calculated at two separate depths. Thus equation (6.24) can be written

$$\Delta_z\left(\bar{u}\frac{\partial\bar{v}}{\partial x}\right) + \Delta_z\left(\bar{v}\frac{\partial\bar{v}}{\partial y}\right) + \Delta_z\left(\bar{w}\frac{\partial\bar{v}}{\partial z}\right) = -\Delta_z\left(\frac{\partial\overline{D}}{\partial y}\right) - \Delta_z(f_1\bar{u}) \qquad (6.25)$$

where Δ_z is the operator $\partial/\partial z$. The assumption of a linear variation of shearing stress with depth is reasonable in the near surface layer where conditions are fairly uniform. Providing the radius of curvature of the streamlines is constant between the two depths, both sides of equation (6.25) should equal the contribution due to the centrifugal force.

Vellar Estuary

From the observations at the surface and at 0·5 m depth, for the periods 20th January, 9th February and 15th February 1967, at Stations 5–10, the lateral dynamic balance in the Vellar Estuary has been evaluated.

The acceleration terms, the pressure term and the Coriolis term in equation (6.24) were calculated at 0·5 m depth. At the first period fresh water occupied the whole of the surface layer throughout the tidal cycle. The largest terms were the contributions of the longitudinal and lateral flows to the lateral acceleration, and they were too large to be balanced by the Coriolis force. The radii of curvature of the streamlines were calculated using equation

(6.25). The acceleration terms gave values less than one kilometre, but balancing Coriolis force against the centrifugal force gave radii between 7 and 30 km. As these are more reasonable it appears that errors in the acceleration terms were probably large at that period. Certainly the data from which the mean flows were calculated were not as complete as at the later periods.

During the other two periods of observation the pressure force due to the observed density distribution was the largest term and was not balanced by the accelerations and Coriolis force. The largest acceleration term was generally the contribution of the lateral flow. The longitudinal and vertical contributions were smaller and of the same order. Using equation (6.25), equating the acceleration terms and the centrifugal force gave streamlines with radii of curvature between 1 and 42 km. Considering the estuary topography the flow patterns were realistic. Equating centrifugal force against the pressure and Coriolis forces gave much smaller values and variations in the direction of the centre of curvature which produced unrealistic flow patterns. Because tidal variations in salinity of up to 25‰ occurred near the surface, the consequent errors in the mean salinities may have produced appreciable errors in the pressure terms. This brings into question the results of the salt balance analysis carried out in Chapter 5. Also, during the final period the level of no net flow may have lain between the surface and 0·5 m depth on the lower section. The values for the numerator of the centrifugal force were mainly governed by the vertical change in the amplitude of the tidal currents.

Fjords

In fjords analysis is often made by assuming a lateral balance between Coriolis force and the pressure gradients.

$$f_1 \bar{u} = -\left(\frac{1}{\bar{\rho}} \frac{\partial \bar{p}}{\partial y}\right) \tag{6.26}$$

The lateral accelerations, the frictional stresses and the body forces are considered negligible. Providing some assumption can be made about the depth of no motion, the depth of a surface with zero horizontal pressure gradients, equation (6.26) can be solved. Cameron (1951) from observations at the entrance to Portland Inlet, British Columbia, used a depth of no motion of 27·4 m and obtained satisfactory agreement between the calculated and observed fresh water discharge.

This process appears to be valid in other straight, deep fjords. Tully (1958) has examined the transports in the Juan de Fuca Strait with the same basic assumptions. The fresh water fraction f of the total seaward volume transport Q_1 in the upper zone equals the river flow.

$$fQ_1 = R$$

The landward transport Q_2 of sea water in the lower layer

$$Q_2 = (1-f)Q_1 = \frac{1-f}{f}R$$

The lateral balance gives

$$\frac{\partial \overline{D}}{\partial y} = f_1(\bar{u}_1 - \bar{u}_2)$$

As

$$\bar{u}_1 - \bar{u}_2 = \frac{Q_1}{A_1} - \frac{Q_2}{A_2}$$

Then

$$Q_1 = \frac{\partial \overline{D}}{f_1 \, \partial y} \frac{A_1 A_2}{A_2 + A_1(1-f)} \quad \text{and} \quad Q_2 = -\frac{\partial \overline{D}}{f_1 \, \partial y} \frac{A_1 A_2 (1-f)}{A_2 + A_1(1-f)}$$

In this case the depth of no motion, to which the dynamic height anomalies were computed, and the depth for separation of the areas A_1 and A_2 were obtained from the salinity distribution. A plot of salinity against logarithmic depth gave three straight lines; upper and lower almost isohaline zones were separated by a halocline. The junction between the halocline and the lower layer was remarkably constant and it was argued that this must be a level where transport only takes place vertically. The dynamic depth anomalies were consequently related to this depth. The salinity at this depth was called the index salinity and was the base salinity used for calculation of the fresh water fraction in the upper layer.

This method of analysis could be used indirectly to measure the tidal velocities, as Tully also pointed out that in many cases the velocity gradient remains nearly constant throughout the tidal cycle. The transverse slope of the pressure surfaces must then be almost constant and instantaneous values as well as mean values could be used in equation (6.26).

Southampton Water

In Chapter 4 the anomalous salinity distribution in Southampton Water was described, the fresher water on the left-hand side probably being caused by the abrupt widening of the estuary at the confluence of the rivers Test and Itchen. The mean surface discharge from the Test flows across to the left-hand side of the estuary to join that issuing from the Itchen. This convergence of water would cause downward velocities on the left-hand side, and upward velocities on the right. The level of no residual motion would be closer to the surface on the right-hand side. If the river discharge were increased, the mean flow would be enhanced and the lateral salinity gradient would increase. As a result the lateral accelerations can be expected to be large just below the confluence.

From the survey data of the 9th–10th August 1966 at four stations on each of two sections, the various terms in equation (6.24) were calculated. On both sections the largest terms were the contribution of the lateral flow to the lateral acceleration, the pressure term and the Coriolis term. To a great extent the pressure and Coriolis terms cancelled.

TABLE 6.5

Lateral Dynamic Balance in Southampton Water

Station no.	$\Delta_z\left(\overline{u}\dfrac{\partial\overline{v}}{\partial x}\right)$	$\Delta_z\left(\overline{w}\dfrac{\partial\overline{v}}{\partial z}\right)$	$\Delta_z\left(\overline{v}\dfrac{\partial\overline{v}}{\partial y}\right)$	$\Delta_z\left(\dfrac{\partial\overline{D}}{\partial y}\right)$	$\Delta_z(f_1\overline{u})$	$\Delta_z(\overline{u^2}+\overline{U^2})$ ×10³ m²	r_{xy}
			×10⁶ m sec⁻²			sec⁻²	km
A 1–2	0·23	0·60	−4·23	5·00	−1·70	−26·0	7·6 7·8
2–3	−1·30	0·47	−0·82	1·37	−4·39	−42·5	25·7 −14·1
3–4	3·14	−2·16	2·92	0·19	−3·88	−17·3	−4·4 −4·7
B 1–2	−0·46	1·07	8·56	9·83	2·26	−3·3	−0·4 0·3
2–3	0·27	0·69	−10·00	9·86	0·46	−6·3	0·7 0·6
3–4	−2·87	1·92	−5·47	12·95	2·37	22·6	3·5 −1·5

Each term in equation (6.25) was calculated for depths of 60 and 200 cm and the results are shown in Table 6.5. Only in three instances do the accelerations and forces nearly balance. In the other cases large remainders are left, indicating large errors somewhere. The radii of curvature were calculated by assuming equivalence of the centrifugal force in turn with the accelerations and forces, on the left and right side of equation (6.25). The results are shown in the final two columns of Table 6.5. Two results on the lower section show similar values with a realistic radius of curvature. One result on the upper section shows similar values, but the radius of curvature is unrealistically small. On this section the mean velocities suggest that the level pressure surface may have lain between 60 and 200 cm depth on the right-hand side. This would give large values for the pressure term and reduce the value of the numerator in the centrifugal force. The values for the numerator, however, are derived mainly from the change in the amplitude of the tidal velocities with depth, rather than vertical changes in the mean velocities. The positioning of the stations obliquely to the probable flow direction would also introduce errors. As each section was observed on different days errors would be introduced if non-steady-state conditions existed.

There are probably very large errors in the lateral accelerations because of errors in the measured lateral velocities. As these are calculated by resolving the components of the measured velocities, they can be affected by small errors in the observations of flow direction. Averaging over a tidal cycle may not be sufficient to remove these errors and spurious lateral mean velocities can be produced. Though lateral accelerations are obviously important in estuaries they are difficult to measure sufficiently accurately to enable a proper dynamic analysis to be completed.

Because of the possible errors in the lateral velocities that have been shown by this analysis, reconsideration of the salt balance, as carried out in Chapter 5, is necessary.

James River

For the James River, Pritchard (1956) has considered that the lateral mean and tidal motions are zero and that the horizontal contributions to the turbulent flux of momentum are negligible. Consequently for steady-state conditions equation (6.21) becomes:

$$0 = -\frac{\partial \overline{D}}{\partial y} + C_2 + f_1 \bar{u} - \frac{\partial}{\partial z}(\overline{v'w'}) \tag{6.27}$$

This equation was evaluated using the two boundary conditions of $(\overline{v'w'}) = 0$ at the sea surface and sea bottom. It revealed that at about mid-depth there was a lateral surface of zero pressure gradient, though this was not necessarily the same as that derived from the longitudinal balance. Above this depth the pressure surfaces sloped downwards towards the left and below this depth they sloped downwards towards the right-hand side. The assumed boundary conditions would not be true in the presence of a surface wind stress or secondary currents.

TABLE 6.6

Magnitudes of the Terms in the Lateral Dynamic Balance of the James River [a]

Depth (m)	$(-\alpha(\partial p/\partial y))$	$f_1 \bar{u}$ m sec$^{-2} \times 10^6$	$\partial/\partial z(\overline{v'w'})$
0	−16·49	10·18	−6·31
1	−11·64	6·28	−5·41
2	−6·26	3·07	−3·19
3	−0·53	−1·23	−1·76
4	5·08	−4·92	0·16
5	9·73	−6·15	3·58
6	12·63	−6·67	5·96
7	13·19	−5·97	7·32

[a] Reproduced with permission from D. W. Pritchard, *Jour. Marine Res.*, **15**, 1956, Table 3.

The magnitudes of the various terms in equation (6.27) are shown in Table 6.6. This shows that the Coriolis force resulting from the mean horizontal motion is balanced mainly by the lateral pressure force. The eddy frictional term is of secondary importance.

Stewart (1958) has questioned the reality of these finite values of $(\overline{v'w'})$ and showed that the lateral balance could be maintained by centrifugal forces without any eddy friction contribution. Stewart showed that the necessary curvature of the streamlines was consistent with the topography of the James River, both at the surface and near the bottom, and that this effect was produced mainly by the fluctuating tidal currents rather than the mean flow.

Following the analysis of the longitudinal and lateral dynamic balances, Pritchard and Kent (1956) suggested that, as the variation with depth of the vertical and lateral shearing stresses were similar, they could be represented as $(\overline{v'w'}) = \eta(\overline{u'w'})$ where η had a value of about 0·4. Then, assuming that the mean field acceleration terms were zero, they were able to use the measured salinities and temperatures to calculate the relative pressure gradient distribution. The field change in the amplitude of the tidal velocity

$$U_o \frac{\partial U_o}{\partial x}$$

was estimated from tidal current data and the mean longitudinal velocity $\bar{u}$ was calculated. The agreement between the calculated and observed values of $\bar{u}$ was good. This illustrates the internal consistency of the data, but, in view of the importance of centrifugal forces, does not really represent the actual situation.

Solution of the general equations of motion in estuaries is more difficult than for the equation of salt continuity. In the latter case we were dealing with the product of a vector and a scalar, whereas with the former we have the product of two vectors. As a consequence we have to be more accurate with our direction measurements and more stringent about our averaging criteria. To diminish the effects of errors in the current direction what is really required is that the x co-ordinate should be directed along the current flow, rather than along the estuary, and, as we have seen, locally the two may be completely different. There can be drastic variations of flow direction during a tide, the flood tide need not be 180° different in direction from the ebb tide. This effect can also alter with depth and will be especially apparent with the residual flows, as the control on the surface mean flow is mainly the upstream topography and on the bottom flow the topography downstream. Therefore, unless careful measurements are made, directional errors can produce rather large errors in the lateral velocities and these will be exaggerated in the calculation of the vertical velocities. Because of the importance of topographic controls on the flow in the lateral direction, estimation of the turbulent stresses is virtually impossible except by direct measurement. Consequently the methods of analysis discussed here probably have little future. There is a great future however, for mathematical models in which real topography can be used. Many present-day mathematical models are two-dimensional models with rather idealized topographies, but the trend is towards the larger, more complex, yet more realistic ones.

CHAPTER 7

Flushing and Pollution Distribution Prediction

FLUSHING TIME

It is obvious from what we have already seen of estuaries, that increased river flow causes both a downstream movement of the salinity intrusion and a more rapid circulation of water. Thus increased river discharge is accompanied by a more rapid exchange of fresh water with the sea, the volume of fresh water accumulated in the estuary increasing to a lesser extent than does the discharge. The flushing time is the time required to replace the existing fresh water in the estuary at a rate equal to the river discharge.

The flushing time $T = Q/R$ where Q is the total amount of river water accumulated in the whole or a section of the estuary and R is the river flow. In Boston Harbour the flushing time changes rapidly with discharge variation at low river flow, but changes slowly at high river flow (Ketchum 1952).

The flushing time can be calculated in several ways.

The Fraction of Fresh Water Method

The mean fractional fresh water concentration over any segment is

$$f = \frac{S_s - S_n}{S_s} \tag{7.1}$$

where S_s is the salinity of the undiluted sea water and S_n is the mean salinity in a given segment of the estuary. The total volume Q is found by multiplying the fractional fresh water concentration 'f' by the volume of the estuary segment. Hughes (1958) has calculated the flushing time of the Mersey Narrows by this method and at a river discharge of 25·7 m³ sec⁻¹ the flushing time was 5·3 days. Large river flows decreased the flushing time.

Ketchum (1950) has used this method in New York Bight where he found that the flushing time varied between 6·0 and 10·6 days with the river flow varying between 4·9 and 0·46 × 10⁹ ft³ day⁻¹.

The Tidal Prism Method

In this method, the water entering on the flood tide is assumed to become fully mixed with that inside and the volume of sea water and river water introduced equals the volume of the tidal prism, the volume between high and low tide marks. On the ebb the same volume of water is removed and the fresh water content of it must equal the increment of the river flow. If V is the low tide volume and P the intertidal volume (the tidal prism) then the flushing time in tidal cycles:

$$T = \frac{V+P}{P} \tag{7.2}$$

It has been found that the flushing time calculated this way gives a considerably lower flushing time than calculation using other methods. The exaggerated estimate of the rate of flushing is due to incomplete mixing of the estuarine water; the fresher water near the head of the estuary cannot reach the mouth during the ebb. Also, some of the water which does escape during the ebb returns on the following flood tide.

The Modified Tidal Prism Method

Ketchum (1951) modified the tidal prism approach by dividing the estuary into segments, the lengths of which are determined by the excursion of a water particle during the tide. The innermost section is that above which the intertidal volume P_o is supplied by the river flow R. Thus $P_o = R$. The low tide volume of this innermost segment is V_o. The limit of the next segment is placed so that $V_1 = V_o + P_o = V_o + R$ etc. The low tide volume in each segment equals the total tidal prism within the next segment to landward, plus the low tide volume in segment O, or $V_n = V_o + R + \sum_1^{n-1} P$.

Each segment contains, at high tide, the volume of water contained in the next seaward segment at low tide. Thus the limits of the segments are equal to the average excursion of a particle of water on the flood tide. If the mixing is complete at high tide then the proportion of water removed on the ebb tide is the ratio between the local intertidal volume and the high tide volume. Thus an exchange ratio can be defined for any segment n as $r_n = P_n/(P_n + V_n)$. The flushing time in tidal cycles will be $1/r_n$.

Providing the river flow is constant, each segment receives R volume of river water per tidal cycle. The amount of river water removed on the ebb will be $r_n R$ and the amount remaining will be $(1-r_n)R$. As this process will have already been going on for many tidal cycles, there will be contributions from the river flow at those times, both to the water removed and to that remaining. This can be summarized as follows:

Age in tidal cycles	River water removed	River water remaining
1	$r_n R$	$(1-r_n)R$
2	$r_n(1-r_n)R$	$(1-r_n)^2 R$
3	$r_n(1-r_n)^2 R$	$(1-r_n)^3 R$
m	$r_n(1-r_n)^{m-1}R$	$(1-r_n)^m R$

The total volume of river water Q_n accumulated in the segment n will be the sum of the last column, plus one volume of river flow which has not yet been removed.

$$Q_n = R(1+(1-r_n)+(1-r_n)^2 \ldots (1-r_n)^m)$$

This is a geometrical progression whose sum is:

$$\frac{R}{r_n}(1-(1-r_n)^{m+1})$$

When m is large, $(1-r_n)^{m+1}$ approaches zero when r_n is less than unity, so that

$$Q_n = R/r_n$$

Similarly the amount of river water removed is:

$$R(r_n+r_n(1-r_n)+r_n(1-r_n)^2 \ldots r_n(1-r_n)^{m-1}) = R$$

Consequently we can calculate the flushing time $(1/r_n)$ for any section and, if we know the undiluted sea water salinity, the high water salinity in each segment can also be calculated. For incomplete mixing at high tide, the exchange ratio can be adjusted by multiplying by $\bar{h}/\overline{H}$ where $\bar{h}$ is the average depth of the segment and $\overline{H}$ is the average depth of the mixed layer. The total flushing time for the estuary will be the sum of the flushing times for the separate segments.

This is a useful method of calculating the flushing time and the salinity distribution, as it only requires knowledge of the river flow, tidal range and the estuarine topography. Also, since the segment length is equivalent to the tidal excursion, some idea of the current velocities within the estuary can be obtained.

The modified tidal prism method has been applied to a number of estuaries with reasonably good agreement to the observed salinity distribution. For other estuaries, such as the Severn (Stommel 1953a) the agreement is not good. It appears that the best results are obtained when the number of segments is large, when the estuarine cross-sectional area increases fairly quickly downstream and when the estuary is well mixed.

Ketchum and Keen (1953) have applied the method to the Bay of Fundy. The total flushing time for the Bay was 76 days, agreeing well with the computed accumulation of fresh water derived from salinity measurements. The segmentation length agreed well with the observed particle excursions and the exchange ratio also agreed well with that calculated from the observed fresh water fraction.

The salinity distribution and segmentation calculated for Southampton Water by Ketchum's method are shown in Figure 7.1. River flows of $31{\cdot}3 \times 10^6$ cu ft per tide and $15{\cdot}7 \times 10^6$ cu ft per tide were used for the Test and Itchen respectively, with a spring tidal range of 15 ft. Comparison with the measured values (Figure 4.24) is difficult because of the effect of Marchwood Power

Station, which tends to bypass some of the surface water flow and create a fresh water depleted zone between the cooling water intake and outfall.

The results of flushing times calculated both by the fractional fresh water method and the modified tidal prism method for the Columbia River have been presented by Neal (1966). The first method gave flushing times of between 1–5 days depending on the river flow and the second method gave about twice these times. The variation of flushing time with river flow for both methods followed very similar trends, however.

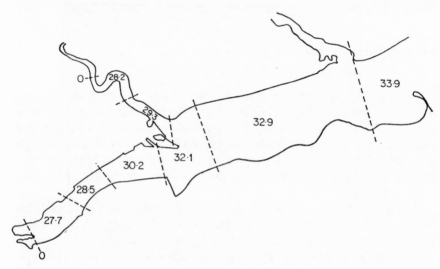

Figure 7.1   Tidal prism segmentation of Southampton Water and calculated high water mean salinities

Theory of a Mixing Length

Using Ketchum's idea that the element of mixing volume is bounded by the length of the tidal excursion, Arons and Stommel (1951) developed a mixing length theory and attempted to produce flushing numbers to characterize estuaries. If the estuary is of uniform width and of uniform depth (h) and the tide is simultaneous and uniform over the entire channel, then the amplitude of the tidal current U_o is $U_o = A_o \omega(x/h)$, where A_o is the amplitude of the vertical tidal movement, x is the distance from the head of the estuary and ω is the angular frequency of the tide ($\omega = 2\pi/T$ where T is the tidal period). The amplitude of the horizontal tidal displacement is

$$\xi_o = -\frac{A_o x}{h}$$

The horizontal eddy–diffusion coefficient K_x is assumed to be related to the tidal displacement and the current by $K_x = 2B\xi_o U_o$, where B is a constant.

Thus

$$K_x = 2BA_o^2 \omega \frac{x^2}{h^2}$$

Introducing a dimensionless length parameter $\lambda = x/L$ where L is the total estuary length, and a flushing number

$$F = \frac{\bar{u}h^2}{2BA_o^2 \omega L}$$

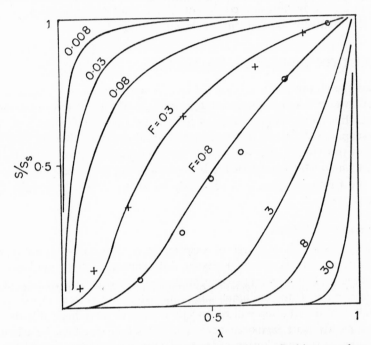

Figure 7.2 Curves showing relationship between flushing number, salinity and length. ○ Raritan River; + Alberni Inlet. (Reproduced with permission from A. B. Arons and H. Stommel, *Transactions, American Geophysical Union*, **32**, No. 3, 421, Figure 1, 1951)

where $\bar{u}$ is the mean velocity of water in the channel due to the river flow, then the one-dimensional diffusion equation (5.26) becomes

$$F\bar{s} = \lambda^2 \, d\bar{s}/d\lambda \qquad (7.3)$$

Integrating; $\log_e \bar{s} = -F/\lambda + F + \log_e S_s$ where S_s is the undiluted sea salinity.

Thus

$$\bar{s}/S_s = e^{F(1-1/\lambda)} \qquad (7.4)$$

The curves of $\bar{s}/S_s$ against λ for various values of F are shown in Figure 7.2. They show a toe near $\lambda = 0$ and a point of inflexion, similar to the normal

5

picture of salinity distribution. The curves are most sensitive to F in the region $0.1 > F > 10$. Arons and Stommel calculated the values of the constant B by comparison with the observed salinity distributions in the Raritan River and Alberni Inlet. Unfortunately, there was an order of magnitude difference in the results, which means that the method fails to predict adequately the intensity of mixing from the given parameters of the estuaries. This may partly be due to the rather unnatural topographic constraints assumed in the analysis, and partly due to the assumed forms of the flushing number and eddy–diffusion coefficient. These were probably chosen to produce the relatively simple form for the equation (7.3).

POLLUTION DISPERSION PREDICTION

Near the mouth of the estuary the fresh water fraction is relatively low as the salt water is only slightly diluted. Enough of the mixture must escape on each tide to remove a volume of fresh water equivalent to the river flow. The escaping volume can thus be an order or more greater than the river flow and it is this volume that is available for the dilution and removal of pollutants. Consequently, estuaries are better at diluting and removing pollution than the tributary river. It is obviously useful to be able to predict these effects.

Conservative Pollutants

If a constant rate discharge of a conservative, non-decaying, pollutant is made into an estuary the tidal mixing will distribute it both upstream and downstream. The maximum concentration will be in the vicinity of the discharge point. If the pollutant acts in the same way as fresh or salt water, the pollutant distribution will be directly related to the salinity distribution, once a steady state has been achieved. Prediction can thus be based on knowledge of the distribution of fresh water in the estuary.

Ketchum (1955) has developed a fractional fresh water method for predicting the concentration of a pollutant. Let the cross-sectional average concentration at the outfall after steady-state conditions have been achieved be C_o. Then

$$C_o = \frac{P}{R} f_o$$

where P is the rate of supply of the pollutant, R is the river discharge and f_o the cross-sectional fractional fresh water concentration.

Downstream of the outfall the pollutant must pass through a cross-section at the same rate as it is discharged at the outfall

$$C_x = C_o \frac{f_x}{f_o} = \frac{P}{R} f_x \qquad (7.5)$$

Upstream of the outfall, the quantity of pollutant carried upstream with the saline water will balance that carried downstream by the mean flow. Its distribution will be directly proportional to the distribution of salinity, inversely proportional to the fresh water fraction.

Thus

$$C_x = C_o \frac{S_x}{S_o} \qquad (7.6)$$

The pollutant distribution will be of the form shown in Figure 7.3. Downstream it will have the same form as the salinity distribution, and upstream

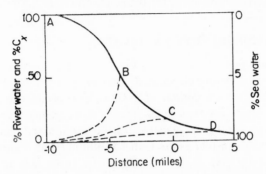

Figure 7.3 Steady-state distribution of a conservative pollutant. (Reproduced with permission from B. H. Ketchum, 1952, *Sewage and Industrial Wastes*, **27**, 1288–1296, Figure 1, Journal Water Pollution Control Federation, Washington, D.C. 20016)

the inverse of the salinity. It is noticeable that if the discharge point is moved downstream, the concentration levels at points seaward of the outfall are unaffected, but the upstream levels are drastically reduced. The concentration in the immediate vicinity of the outfall is also reduced.

Pritchard (1969) has produced a two-dimensional box model for predicting pollutant distribution which is especially applicable to partially mixed estuaries. He considers the estuary to be divided longitudinally into a number of segments, each of which is divided vertically into two sub-segments. The interface between the vertical segments is the boundary between the seaward flowing surface layer and the landward flowing bottom layer. The continuity of salt and water between each box is then considered assuming negligible longitudinal diffusion.

Refering to Figure 7.4, let $(Q_u)_{n-1,n}$ be the volume flow rate from the $n-1$th segment into the nth segment, upper layer and $(Q_L)_{n,n-1}$ the flow in the lower layer from the nth to the $n-1$th segment. Similarly $(Q_u)_{n,n+1}$ will be the upper layer flow rate from the nth to the $n+1$th segment, etc. Within the segment n there will be a volume rate of flow due to vertical advection from the lower into the upper layer (Q_v) and a vertical exchange coefficient representing the vertical diffusion, E_n. The salinity of the upper and lower layers is $(S_u)_n$ and $(S_L)_n$ respectively and the salinity at the boundary between them is $(S_v)_n$. The salinities at the other boundaries will be $(S_u)_{n-1,n}$, $(S_L)_{n+1,n}$ etc. Homogeneity is assumed within each box.

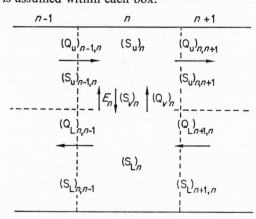

Figure 7.4 Definition diagram for two-dimensional box model, after Pritchard (1969)

For a steady-state salt distribution, for the upper layer in segment n,

$$(S_u)_{n,n+1} \cdot (Q_u)_{n,n+1} = (S_u)_{n-1,n} \cdot (Q_u)_{n-1,n} + E_n[(S_L)_n - (S_u)_n] +$$
$$(Q_v)_n \cdot (S_v)_n \qquad (7.7)$$

From volume continuity:

$$(Q_v)_n = (Q_u)_{n,n+1} - (Q_u)_{n-1,n} \qquad (7.8)$$

There are similar equations for the lower layer. Similar to equation (5.20), for any section of the estuary,

$$Q_u = R\frac{S_L}{S_L - S_u} \qquad Q_L = R\frac{S_u}{S_L - S_u} \qquad (7.9)$$

If the salinity distribution and fresh water flow are known the horizontal volume flow rates can be calculated using equation (7.9). The vertical flow rates can be calculated from equation (7.2) and then equation (7.7) solved for E_n.

Now let a conservative pollutant be introduced into the surface layer of the kth segment at a rate P. The concentration C of the pollutant is assumed uniform in each sub-segment and at the boundaries the concentration is equal to the average value between adjacent segments. Then for the upper layer the pollutant balance is:

$$(Q_u)_{n,n+1}\left[\frac{(C_u)_n + (C_u)_{n+1}}{2}\right] = (Q_u)_{n-1,n}\left[\frac{(C_u)_{n-1} + (C_u)_n}{2}\right]$$
$$+ E_n[(C_L)_n - (C_u)_n] + (Q_v)_n\left[\frac{(C_u)_n + (C_L)_n}{2}\right] \qquad (7.10)$$

For the lower layer:

$$(Q_L)_{n,n+1}\left[\frac{(C_L)_n+(C_L)_{n+1}}{2}\right] = (Q_L)_{n,n-1}\left[\frac{(C_L)_{n-1}+(C_L)_n}{2}\right]$$
$$+(Q_v)_n\left[\frac{(C_L)_n+(C_u)_n}{2}\right]+E_n[(C_L)_n-(C_u)_n] \qquad (7.11)$$

In the upper layer of the kth segment there is the input term P to be added to equation (7.10).

Using equation (7.8), equation (7.10) and (7.11) become:

$$(C_u)_{n-1}\cdot(Q_u)_{n-1,n}-2(C_u)_nE_n+(C_L)_n[2E_n+(Q_v)_n]-$$
$$(C_u)_{n+1}(Q_u)_{n,n+1} = 0 \qquad (7.12)$$

and

$$(C_L)_{n+1}(Q_L)_{n+1,n}-2(C_L)_nE_n+(C_u)_n[2E_n-(Q_v)_n]-$$
$$(C_L)_{n-1}(Q_L)_{n,n-1} = 0 \qquad (7.13)$$

In the kth segment equation (7.12) will again have the additional term P.

Using the boundary conditions that upstream C goes to zero and S goes to zero and downstream that $(C_L)_m = 0$ these equations can be solved for the distribution of pollutant concentration for a given input, with the inherent assumption that the vertical exchange coefficient is the same for pollutant as for salt.

An alternative method of calculating the distribution is by use of the one-dimensional diffusion equation (Stommel 1953a). The net seaward flux of pollutant through any section x is

$$F(x) = Rc-\overline{A}K_x\frac{dc}{dx} \qquad (7.14)$$

where R is the river discharge.

Downstream of the source the net flux must be constant and equal to the input. Upstream it will be zero. Providing the diffusion coefficient for the pollutant can be assumed to be the same as that for salt, then we can determine K_x by putting $F(x)$ equal to the river flow and the fresh water fraction f for c. Then

$$K_x = \frac{R(f-1)}{\overline{A}\ df/dx} \qquad (7.15)$$

The values of K_x calculated this way are put into equation (7.14) and the equation is written in a finite difference form and is solved by successive approximation. There are various constraints on the solution; the concentration must approach zero at the ocean and at the head of the estuary and, in the section near the outfall, the difference in the flux upstream and downstream must equal the rate of inflow of pollutant at the source. According to Neal (1966), the methods of Ketchum and Stommel give similar results.

Pollutant dispersion in a one-dimensional estuary has also been considered by Kent (1960). Kent considers a sectionally homogeneous estuary and

$$\frac{\partial \bar{s}}{\partial t} = -\bar{u}\frac{\partial \bar{s}}{\partial x} + \frac{1}{A}\frac{\partial}{\partial x}\left(AK_q\frac{\partial \bar{s}}{\partial x}\right) \tag{7.16}$$

where K_q is the eddy–diffusion coefficient for the pollutant.

In the first instance a solution for equation (7.16) is considered with values of u, A and K_q variable with x. The equation can be written in a finite difference form and, provided the initial distribution of the pollutant in the estuary is known, we can trace the disposition of the pollutant in space and time. This is done by solving the finite difference equation numerically by successive approximations. To make sure the approximate solution converges on the exact one, certain conditions are specified for the increments of time and distance used for the differences.

A solution is also considered with constant coefficients. In this case equation (7.16) reduces to:

$$\frac{\partial \bar{s}}{\partial t} = -\bar{u}\frac{\partial \bar{s}}{\partial x} + K_q\frac{\partial^2 \bar{s}}{\partial x^2} \tag{7.17}$$

This equation can also be stated in a difference form and solved numerically by successive approximations. Conditions for the space–time increments are again specified to provide convergence of the solution.

In both these analyses it is necessary to know the distribution of the pollutant diffusion coefficient K_q. The diffusion coefficient for salt can be calculated in the same manner as already described, by use of equation (7.15). Kent (1960) modifies this value of K_x, to obtain K_q by arguing that the ratio of diffusivities will be the same as the ratio of the lengths of the salt and pollutant extents. Thus

$$\frac{K_q}{K_x} = \frac{L_q}{L_s}$$

Kent's method requires knowing the initial pollutant distribution shortly after discharge commencement, but experiments with dye releases into an hydraulic model of the Delaware Estuary showed good comparison between observed and predicted distributions thereafter.

Preddy (1954) has modelled pollution in the Thames using a mixing concept not unlike that of Ketchum. He considers that after a tidal cycle a proportion P_1 of the pollutant will be dispersed over a length L downstream, and a proportion P_2 over the same length upstream. The amount $1 - P_1 - P_2$ will be at the discharge point. To determine the pollutant distribution it is necessary to determine the values for P_1 and P_2 at different points in the estuary. By considering the continuity of salt and of water, two equations can be formed in which L, P_1 and P_2 are unknowns. The river flow and salinity distribution need to be known and the length L is taken as the tidal excursion. The equations can then be solved by successive approximations. Preddy uses this method to predict the effect on the salinity distribution of changes in river flow.

Non-conservative Pollutants

For non-conservative pollutants prediction becomes more difficult. Concentrations, as well as diminishing because of mixing, also decrease with time. For coliform sewage bacteria the change due to mortality is an exponential effect that can be represented by

$$C_t = C_o\, e^{kt}$$

The constant k is negative and the population decreases to $1/10$ in $1\frac{1}{2}$ days to 3 days (Ketchum, 1955). As this time is comparable to the flushing time mortality will be important. In the Raritan River practically all of the decrease in bacterial concentration was due to mixing, mortality and grazing by zooplankton. These effects are assumed to be represented by

$$C_n = (C_o)_n \frac{r_n}{1-(1-r_n)e^k} \tag{7.18}$$

where r_n is the exchange ratio in segment n.

If the mortality rates and exchange ratios are equal in the segments in the estuary, then downstream of the outfall

$$C_n = C_o \frac{f_n}{f_o}\left(\frac{r}{1-(1-r)e^k}\right)^n \tag{7.19}$$

and upstream

$$C_n = C_o \frac{S_n}{S_o}\left(\frac{r}{1-(1-r)e^k}\right)^n \tag{7.20}$$

where n is the number of segments from the outfall, the segments being defined by a modified tidal prism analysis. If the coefficient of mortality is zero, i.e. $e^k = 1$, then these equations equal those for a conservative pollutant. The effect of mortality is shown by

$$\frac{C_n}{C_x} = \left(\frac{r}{1-(1-r)e^k}\right)^n \tag{7.21}$$

Larger exchange rates give larger populations in a segment for a given mortality, since the water is mixed faster and less decay occurs. Figure 7.5 shows the distribution of non-conservative pollutant in Raritan Bay where the exchange ratio is 0·34. The upper curve shows the effects of dilution alone, with a conservative pollutant discharged from position A. The lower three curves are for the fraction of the population expected when $k = -0·578$ tide^{-1}. Multiplication of the value in the upper curve by that in the lower gives the expected concentration, relative to a population size of unity for the pollutant mixed in water with no mortality. This example gives relative populations at the outfall locations of 0·538 for A, 0·278 for B and 0·092 for C.

The peak concentration is thus decreased by downstream movement of the outfall, but to a greater extent than for a conservative pollutant. In contrast to a conservative pollutant, concentration at a point downstream of the outfall is reduced if the outfall is moved upstream.

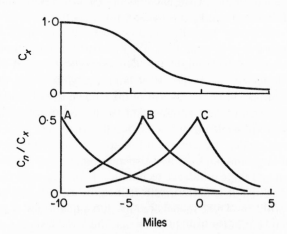

Figure 7.5 Distribution of a non-conservative pollutant in Raritan Bay. For explanation see text. (Reproduced with permission from B. H. Ketchum, 1952, *Sewage and Industrial Wastes*, **27**, 1288–1296, Figure 4, Journal Water Pollution Control Federation, Washington, D.C., 20016)

Stommel (1953a) has also introduced a term into equation (7.14) to allow for decay of pollutant. The equation becomes

$$\frac{d}{dx}\left(Rc - \overline{A}K_x\frac{dc}{dx}\right) + \frac{\overline{A}c}{t} = 0 \qquad (7.22)$$

where t is the time for the concentration to decay to $1/e$ of its initial concentration. This equation can be solved by the same successive approximation method as that used for conservative pollutants.

Many other methods for determining pollutant distributions in estuaries have been published. The relative merits of some of them are discussed by Pyatt (1964).

The consideration of flushing time and the methods discussed for pollution distribution prediction are mathematical models of increasing complexity. Because they use fresh or salt water as the tracer, the results will only apply to substances that act as and are introduced in the same way as either of these.

The methods rely on using the existing salinity distribution and the principles of salt continuity to determine longitudinal dispersion coefficients or exchange coefficients which are then used for the pollutant distribution. These models are then generally tested for reliability by calculating the modelled effect of variation in river discharge and comparing with observed values. Sometimes the mathematical models are tested against observed distributions of dye tracer in hydraulic models. In these cases, however, because of the vertical scale exaggeration in the hydraulic model the dispersion coefficients are likely to be different from those obtained in the prototype estuary.

The most commonly modelled non-conservative pollutant is dissolved oxygen. Though it is not a pollutant, it is a mirror for pollution in which the biological demand for oxygen is large. Consequently a sewage disposal into an estuary will deplete the oxygen levels and to a large extent the dissolved oxygen distribution will be the inverse of the effluent distribution. Modelling the dissolved oxygen content is important so that conditions leading to low oxygen levels and possible anaerobic bacterial activity can be avoided.

Mathematical models are suitable for analysis using computers which can quickly do the large amount of tedious calculation necessary in the successive approximation and finite difference techniques. However, even then, to make the problem tractible it is necessary to reduce the estuary to a one- or two-dimensional problem. These models require large quantities of good quality real data taken simultaneously at many stations over several tidal cycles to determine the initial boundary conditions and for validation. Thus the exchange coefficients are determined empirically rather than analytically.

CHAPTER 8

Theoretical Solutions for Estuaries

As we have seen in chapters 5 and 6, the salt balance and the dynamical balance are parts of the same interlocking system. It would be convenient if it were possible to solve the necessary equations simultaneously. There are a variety of approaches, but one of the most promising is that which has been made by Rattray and Hansen (1962), Hansen and Rattray (1965, 1966) and Hansen (1967). They consider a partially mixed estuary of rectangular cross-section, with breadth b and depth h. The estuary is sufficiently narrow to be laterally homogeneous. In the central portion of such an estuary the vertical salinity stratification is almost independent of position and the longitudinal variation of salinity is almost linear. At either end the stratification is proportional to the departure of the sectional mean salinities from their terminal values, and the salinity distribution is not linear.

Following Pritchard's analyses of the James River, the dynamic balance and the salt balance equations are reduced to:

$$\frac{1}{\rho}\frac{\overline{\partial p}}{\partial x} = \frac{\partial \overline{\tau}_{xz}}{\partial z} = \frac{\partial}{\partial z}\left(N_z\frac{\partial \overline{u}}{\partial z}\right) \tag{8.1}$$

and

$$b\left(\overline{u}\frac{\partial \overline{s}}{\partial x}+\overline{w}\frac{\partial \overline{s}}{\partial z}\right) = \frac{\partial}{\partial x}\left(bK_x\frac{\partial \overline{s}}{\partial x}\right)+\frac{\partial}{\partial z}\left(bK_z\frac{\partial \overline{s}}{\partial z}\right) \tag{8.2}$$

also

$$\frac{1}{\rho}\frac{\overline{\partial p}}{\partial z} = g$$

An equation of state for the water is written as $\rho = \rho_f(1+\kappa s)$ where ρ_f is the fresh water density, and a stream function ψ is introduced into the equation of continuity for water, so that

$$\frac{\partial \psi}{\partial z} = -b\overline{u} \quad \text{and} \quad \frac{\partial \psi}{\partial x} = +b\overline{w}$$

These partial differential equations have no known general solutions, but a group of special solutions can be found when the boundary conditions are

122

appropriate. In such cases ordinary differential equations are obtained from the partial differential equations by transformation of variables. The solutions of the ordinary differential equations are known as similarity solutions.

The boundary conditions that have to be satisfied are: no slip at the bottom, a shearing stress at the sea surface equalling the wind stress, a net transport equalling the river flow and zero salt flux through the boundaries.

For the outer part of an estuary Rattray and Hansen (1962) produce similarity solutions under the conditions that K_x is zero, there is no vertical advection, N_z and K_z and breadth are constant, the net transport due to the river flow is comparatively small and the surface salinity distribution can be described. The analyses gave good comparison between calculated salinity and velocity profiles and those observed in the James River.

The central part of the estuary has been considered by Hansen and Rattray (1965) and Hansen (1967). In this part the vertical exchange coefficients N_z and K_z cannot vary along the estuary, but K_x must increase seaward at a rate equal to the integral mean velocity, or fresh water discharge velocity, i.e.

$$\frac{d}{dx} K_x = u_f \qquad \text{where} \qquad u_f = R/\overline{A}$$

The solutions for the horizontal velocity and salinity distributions are:

$$\frac{u}{u_f} = -\frac{\partial \phi}{\partial \eta} \tag{8.3}$$

and

$$\frac{s}{S_o} = 1 + v\xi + \frac{v}{M}\left[(\eta - \tfrac{1}{2}) - \tfrac{1}{2}(\eta^2 - \tfrac{1}{3}) - \int_0^\eta \phi \, d\eta + \int_0^1 \int_0^\eta \phi \, d\eta' \, d\eta\right]$$

where

$$\phi(\eta) = \tfrac{1}{4}(2 - 3\eta + \eta^3) - \frac{T}{4}(\eta - 2\eta^2 + \eta^3) - \frac{vRa}{48}(\eta - 3\eta^3 + 2\eta^4) \tag{8.5}$$

M is a tidal mixing parameter $= K_z K_{xo} b^2 / R^2$

T is a dimensionless wind stress $= bk^2 \tau_w / N_z R$

Ra is an estuarine Rayleigh number $= gkS_o h^3 / N_z K_{xo}$

S_o is the salinity at $x = 0$

K_{xo} is the horizontal eddy–diffusion coefficient at $x = 0$

$\eta = z/h$

ξ is a dimensionless horizontal co-ordinate $= R_x / bh K_{xo}$

v represents the diffusive fraction of the total upstream salt flux.

Equation (8.5) represents the circulation as the sum of three modes, the river discharge mode, the wind stress mode and the gravitational convection mode associated with the Rayleigh number. The gravitational convection arises from the existence of horizontal salinity differences which attempt to form a horizontal salinity interface by producing a two-layer flow. The less dense water on the surface and towards the head of the estuary is attempting to flow outwards over the salty bottom water that is trying to flow landwards. This has no effect on the net transport of water, which is solely carried on the river discharge mode. However, when coupled with a vertical variation of salinity the gravitational convection does give a net upstream salt flux. This nearly balances the salt advected seawards on the net flow of the river discharge mode. Exchanges of salt by diffusion complete the balance. Because *Ra* is largely related to the total estuary depth, it is possible to have well-developed gravitational convection even though tidal mixing almost creates vertical homogeneity.

With no wind stress the form of the velocity profile depends only on *vRa*, as shown in Figure 8.1. When *vRa* is small the velocity profile tends to the

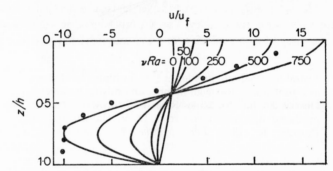

Figure 8.1   Horizontal velocity profile with no wind stress. Observed values (solid dots) for James River Station J–17. (Reproduced with permission from D. V. Hansen and M. Rattray, Jr., 1965, *Jour. Marine Res.*, **23**, 104–122, Figure 3)

parabolic form characteristic of parallel flows with constant viscosity. As the density gradient increases the flow becomes bidirectional for *vRa* > 30. The salinity profile depends on both *M/v* and *vRa* as shown in Figure 8.2.

Generally, there is insufficient information about the exchange coefficients to allow determination of the velocity and salinity profiles directly from equations (8.3)–(8.5). However, approximate values of *vRa* and *M/v* can be obtained by comparison between observed profiles and Figures 8.1 and 8.2. The diffusive fraction *v* can then be obtained by putting these values in equation (8.6).

The diffusive fraction *v* is the positive root of

$$1680M(1-v) = (32+10T+T^2)v+(76+14T)\frac{Rav^2}{48} +\frac{152}{3}\left(\frac{Ra}{48}\right)^2 v^3 \quad (8.6)$$

From the values of *Ra* and *M* and from the observed horizontal salinity gradient, which allows calculation of K_{xo}, values for the other exchange coefficients can be calculated.

Hansen and Rattray (1965) report values of the various parameters and exchange coefficients calculated in this manner from published observations in a number of estuaries. They compare well with results deduced by other means.

The stratification increases with two-layer flow of increasing intensity, i.e. with increased gravitational convection (increasing vRa). The increased

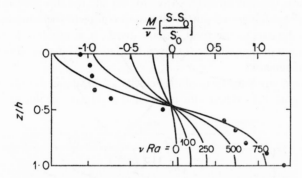

Figure 8.2   Salinity profiles at $\xi = 0$ with no wind stress, as a function of Rayleigh number and mixing parameter. Observed values for James River J–17 for $M/v = 8$. (Reproduced with permission from D. V. Hansen and M. Rattray, Jr., 1965, *Jour. Marine Res.*, **23**, 104–122, Figure 4)

stratification reduces the proportion (v) of salt exchange by diffusion. Thus when v is small, diffusion is negligible and the necessary upstream salt flux is entirely by gravitational convection associated with two-layer flow. When v is near unity, gravitational convection ceases to be important, two-layer flow does not exist and diffusion accounts for the upstream salt transfer. As v increases and decreases respectively as the mixing parameter M and the stability parameter Ra increase, then M/v and vRa change less than M and Ra. Consequently the vertical profiles of salinity and velocity tend to be stabilized, the diffusive fraction forms a buffer preventing excessive changes in the vertical salinity and velocity profiles. Figure 8.3 shows the relationship between v, M and Ra for the special case of zero wind stress.

From results in a number of estuaries Hansen and Rattray (1966) have also indicated that

$$vRa = 16F_m^{-\frac{3}{4}} \quad \text{and} \quad M/v = 0.05 \, P^{-7/5} \tag{8.7}$$

F_m is the densimetric Froude number $F_m = u_f/\sqrt{gh\Delta\rho/\rho}$, $P = u_f/u_t$ where u_t is the root mean square tidal current speed.

Different types of estuaries can be characterized by different values of F_m and P and also by the ratios $\delta S/S_o$, a stratification parameter, and u_s/u_t, a circulation parameter. The importance of these parameters is apparent in

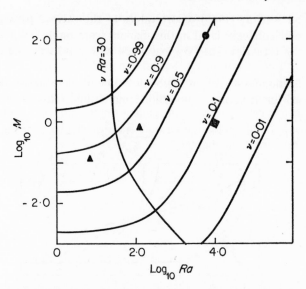

Figure 8.3 Relationship between *M*, *Ra* and *v* given by equation (8.6). ●, Mersey Estuary; ◆, James River; ▲, Columbia River. (Reproduced with permission from D. V. Hansen and M. Rattray, Jr., 1965, *Jour. Marine Res.*, **23**, 104–122, Figure 5)

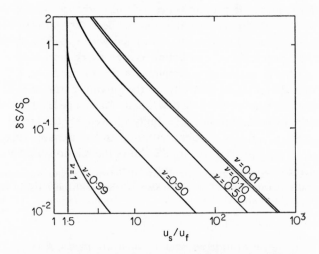

Figure 8.4 Fraction of horizontal salt balance by diffusion, as a function of salinity stratification and convective circulation in a rectangular channel. (Reproduced with permission from D. V. Hansen and M. Rattray, Jr., 1966, *Limnol. Oceanog.*, **11**, 319–326, Figure 1)

equations (8.3) and (8.4) and their relationships with v and with F_m and P are shown in Figures 8.4 and 8.5.

Figure 8.5 shows that the circulation parameter depends solely on F_m, but that the stratification parameter depends on both F_m and P. Fischer (1972) replotted Hansen and Rattray's results in terms of the estuarine Richardson number and densimetric Froude number and found that the stratification parameter then became primarily dependent on Ri_E and only slightly on F_m. The stratification and circulation parameters form the basis of the system of classification described in chapter 2.

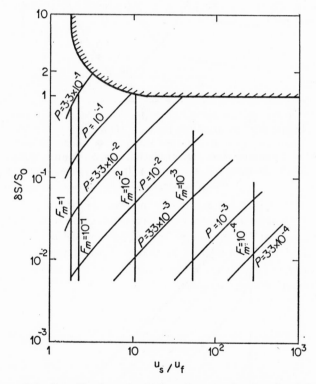

Figure 8.5 Stratification–circulation diagram showing isopleths of the bulk parameters F_m and P. (Reproduced with permission from D. V. Hansen and M. Rattray, Jr., 1966, *Limnol. Oceanog.*, **11**, 319–326, Figure 4)

Values for the stratification and circulation parameters in the Vellar estuary (Dyer and Ramamoorthy, 1969) are shown in Figure 4.10. The values of F_m and P calculated from the observations agree well with those predicted in Figure 8.5 and calculation of v, as discussed in chapter 5, also gives reasonable agreement with the values predicted in Figure 8.4.

The stratification–circulation diagrams for four sections of the Mersey (Bowden and Gilligan, 1971) give lines that tend to follow a constant value of the diffusive fraction v (Figure 4.22). A comparison between the range of values for v predicted from Figure 8.4 and those calculated (chapter 5) are shown in Table 8.1. The agreement is generally good, though the predicted

TABLE 8.1

Diffusive Fraction of Salt Transport v from Direct Calculation and Predicted by the Stratification–Circulation diagram for the Mersey Narrows [a]

Section (see Figure 4.18)	No. of observations	v predicted range	v calculated Mean	S . D
RR	8	0·7–0·9	0·85	0·04
EE	11	0·3–0·55	0·51	0·16
CC	5	0·1–0·25	0·30	0·11
DD	6	0·3–0·6	0·62	0·15

[a] Reproduced with permission from K. F. Bowden and R. M. Gilligan, *Limnol. Oceanog.*, **16**, 490–502, 1971, Table 2.

values tend to be slightly less than those measured. Values of P and F_m were also calculated from the data and, using Hansen and Rattray's relationships, values of the stratification and circulation parameters were calculated from them to compare with those measured. The result of the comparison is shown in Table 8.2. The best agreement is in the centre section of the channel where the dimensions best fit the assumptions of constant width and depth.

The same similarity principles have been used at the inner end of the estuary (Hansen and Rattray, 1965). For this case it is necessary to have the bottom sloping, though the width remains constant. The vertical salinity

TABLE 8.2

Ratio of Measured and Predicted Stratification and Circulation Parameters for the Mersey Narrows [a]

Section (see Figure 4.18)	No. of observa- tions	Circulation parameter measured/predicted Mean	S . D	Stratification parameter measured/predicted Mean	S . D
RR	8	0·42	0·26	0·77	0·16
EE	9	1·05	0·40	1·02	0·51
CC	6	0·97	0·40	2·05	0·93
DD	6	0·95	0·73	1·73	0·85

[a] Reproduced with permission from K. F. Bowden and R. M. Gilligan, *Limnol. Oceanog.*, **16**, 490–502, 1971, Table 3.

profiles for large values of vRa closely resemble those of Figure 8.2 and the longitudinal velocity profiles are the same as those in Figure 8.3. The main difference is that vertical velocities are also required. The vertical velocities are formed of two contributions: one associated with the local addition of river water, the other associated with the bottom slope. The bottom slope contribution leads to a downward velocity in the surface layer and an upward velocity in the bottom layer for $vRa > 30$.

One simple solution at the inner end can be found providing K_x and K_z are independent of x and N_z is proportional to the mean salinity and its gradient. This solution provides results that agree well with data from the James River.

Rattray (1967) has extended the analysis to fjords, but it becomes more complex as a tidal acceleration term has to be included and the depth has to be considered as infinite. However, reasonable agreement with observed data is obtained.

Hansen and Rattray (1972) have applied the similarity technique to inlets where circulation is generated by diffusive modification of stratification maintained by external dynamics, rather than by fresh water flow directly into the inlet. The solutions compare well with observations on Baltimore Harbour and potentially are useful in the interpretation of three-layer flow systems.

In spite of the rather stringent conditions imposed on the solutions, the approach appears to be most promising and bears real correspondence to measured conditions. Eventually, refinement of the method may allow some of the most inappropriate of the assumptions to be relaxed.

References

Arons, A. B. and Stommel, H., 1951. 'A mixing length theory of tidal flushing', *Trans. Amer. Geop. Un.*, **32**, 419–421.

Blanton, J., 1969. 'Energy dissipation in a tidal estuary', *Jour. Geop. Res.*, **74**, 5460–5466.

Bowden, K. F., 1960. 'Circulation and mixing in the Mersey Estuary', *I.A.S.H. Comm. Surface waters*, Publ. 51., 352–360.

Bowden, K. F., 1962. 'Measurements of turbulence near the sea bed in a tidal current', *Jour. Geop. Res.*, **67**, 3181–3186.

Bowden, K. F., 1963. 'The mixing processes in a tidal estuary', *Int. Jour. Air Water Poll.*, **7**, 343–356.

Bowden, K. F., 1965. 'Horizontal mixing in the sea due to a shearing current', *Jour. Fluid Mech.*, **21**, 83–95.

Bowden, K. F. and Fairbairn, L. A., 1956. 'Measurements of turbulent fluctuations and Reynolds stresses in a tidal current', *Proc. Roy. Soc.*, **A237**, 422–438.

Bowden, K. F., Fairbairn, L. A. and Hughes, P., 1959. 'The distribution of shearing stresses in a tidal current', *Geophy. Jour. Roy. Ast. Soc.*, **2**, 288–305.

Bowden, K. F. and Gilligan, R. M., 1971. 'Characteristic features of estuarine circulation as represented in the Mersey Estuary', *Limnol. Oceanog.*, **16**, 490–502.

Bowden, K. F. and Howe, M. R., 1963. 'Observations of turbulence in a tidal channel', *Jour. Fluid Mech.*, **17**, 271–284.

Bowden, K. F. and Proudman, J., 1949. 'Observations on the turbulent fluctuations of a tidal current', *Proc. Roy. Soc.*, **A199**, 311–327.

Bowden, K. F. and Sharaf el Din, S. H., 1966a. 'Circulation, salinity and river discharge in the Mersey Estuary', *Geophy. Jour. Roy. Ast. Soc.*, **10**, 383–400.

Bowden, K. F. and Sharaf el Din, S. H., 1966b. 'Circulation and mixing processes in the Liverpool Bay area of the Irish Sea', *Geophy. Jour. Roy. Ast. Soc.*, **11**, 279–292.

Cameron, W. M., 1951. 'On the transverse forces in a British Columbia Inlet', *Trans. Roy. Soc. Canada*, **45**, 1–9.

Cameron, W. M. and Pritchard, D. W., 1963. 'Estuaries', in *The Sea* (ed. M. N. Hill), Vol. 2, John Wiley & Sons, New York, 306–324.

Cannon, G. A., 1971. 'Statistical characteristics of velocity fluctuations at intermediate scales in a coastal plain estuary', *Jour. Geop. Res.*, **76**, 5852–5858.

Carstens, T., 1970. 'Turbulent diffusion and entrainment in two-layer flow', *Proc. Amer. Soc. Civil Eng.*, WW1, 97–104.

Defant, A., 1961. *Physical Oceanography*, Vol. II, Pergamon Press, Oxford.

Duke, C. M., 1961, 'Shoaling of the lower Hudson River', *Proc. Amer. Soc. Civil Eng.*, **87**, WW1, 29–45.

Dyer, K. R., 1972. 'Sedimentation in Estuaries', in *The Estuarine Environment*, (ed. R. S. K. Barnes and J. Green), Applied Science, London, 133p.

Dyer, K. R. and Ramamoorthy, K., 1969. 'Salinity and water circulation in the Vellar Estuary', *Limnol. Oceanog.*, **14**, 4–15.

Einstein, H. A. and Shen, H. W., 1964, 'A study on meandering in straight alluvial channels', *Jour. Geop. Res.*, **69**, 5239–5247.

Elder, J. W., 1959. 'The dispersion of marked fluid in turbulent shear flow', *Jour. Fluid Mech.*, **5**, 544–560.

Farmer, H. G. and Morgan, G. W., 1953, 'The salt wedge', *Proc. 3rd Conf. Coastal Eng.*, 54–64.

Farrell, S. C., 1970. 'Sediment distribution and hydrodynamics Saco River and Scarboro estuaries, Maine', Cont. 6-CRG. Dept. Geol. Univ. Mass.

Fischer, H. B., 1967. 'Analytic prediction of longitudinal dispersion coefficients in natural streams', *Proc. 12th Cong. Int. Ass. Hyd. Res.*

Fischer, H. B., 1972. 'Mass transport mechanisms in partially stratified estuaries', *Jour. Fluid Mech.*, **53**, 672–687.

Grant, H. L., Stewart, R. W. and Moillet, A., 1962. 'Turbulence spectra from a tidal channel', *Jour. Fluid Mech.*, **12**, 241–263.

Hansen, D. V., 1965. 'Currents and mixing in the Columbia River estuary', *Trans. Joint. Conf. Ocean science and Ocean engineering*, 943–955.

Hansen, D. V., 1967. 'Salt balance and circulation in partially mixed estuaries', in *Estuaries* (ed. G. H. Lauff) Amer. Assoc. Adv. Sci.

Hansen, D. V. and Rattray, M. Jr., 1965. 'Gravitational circulation in estuaries', *Jour. Marine Res.*, **23**, 104–122.

Hansen, D. V. and Rattray, M. Jr., 1966. 'New dimensions in estuary classification', *Limnol. Oceanog.*, **11**, 319–326.

Hansen, D. V. and Rattray, M. Jr., 1972, 'Estuarine circulation induced by diffusion', *Jour. Marine Res.*, **30**, 281–294.

Harleman, D. R. F. and Ippen, A. T., 1960. 'The turbulent diffusion and convection of saline water in an idealized estuary', *I.A.S.H. Comm. Surface Water*, Publ. 51, 362–378.

Hughes, P., 1958. 'Tidal mixing in the Narrows of the Mersey Estuary', *Geophy. Jour. Roy. Ast. Soc.*, **1**, 271–283.

Ippen, A. T., 1966. 'Estuary and coastline hydrodynamics', McGraw-Hill, New York.

Ippen, A. T. and Harleman, D. R. F., 1961. 'One-dimensional analysis of salinity intrusion in estuaries', Tech. Bull. 5. Comm. Tidal Hydraul. Corps. Eng. U.S. Army.

Kent, R. E., 1960. 'Diffusion in a sectionally homogeneous estuary', *Proc. Amer. Soc. Civil Eng.*, **86**, SA 2, 15–47.

Ketchum, B. H., 1950. 'Hydrographic factors involved in the dispersion of pollutants introduced into tidal waters', *Jour. Boston Soc. Civil Eng.*, **37**, 296–314.

Ketchum, B. H., 1951. 'The exchanges of fresh and salt water in tidal estuaries', *Jour. Marine Res.*, **10**, 18–38.

Ketchum, B. H., 1952. 'Circulation in estuaries', *Proc. 3rd Conf. Coastal Eng.*, 65–76.

Ketchum, B. H., 1955. 'Distribution of coliform bacteria and other pollutants in tidal estuaries', *Sewage Indust. Wastes*, **27**, 1288–1296.

Ketchum, B. H. and Keen, D. J., 1953. 'The exchanges of fresh and salt waters in the Bay of Fundy and in Passamquoddy Bay', *Jour. Fish. Res. Bd. Canada*, **10**, 97–124.

Keulegan, G. H., 1949. 'Interfacial instability and mixing in stratified flows', *Jour. Res. Nat. Bur. Stds.*, **43**, 487–500.

La Fond, E. C., 1960. 'Processing oceanographic data', U.S. Hydrographic Office, Pub. 614.

Lauff, G. H., 1967. *Estuaries*, Publ. 83, Amer. Assoc. Adv. Sci.

Leopold, L. B. and Wolman, M. G., 1960. 'River meanders', *Bull. Geol. Soc. Amer.*, **71**, 769–794.

Lofquist, K., 1960. 'Flow and stress near an interface between liquids', *Phys. Fluids*, **3**, 158–175.

Macagno, E. O. and Rouse, H., 1962. 'Interfacial mixing in stratified flow', *Trans. Amer. Soc. Civil Eng.*, **127**, 102–128.

McAlister, W. B., Rattray, M. Jr. and Barnes, C. A., 1959. 'The dynamics of a fjord estuary: Silver Bay, Alaska', Tech. Rept. 62, Univ. Washington. Dept. Oceanography.

Neal, V. T., 1966. 'Predicted flushing times and pollution distribution in the Columbia River estuary', 10th Conf. Coastal Eng., 1463–1480.

Okubo, A., 1964. 'Equations describing the diffusion of an introduced pollutant in a one-dimensional estuary', in *Studies on Oceanography* (ed. K. Yoshida), Univ. Washington Press.

Okubo, A., 1971. 'Oceanic diffusion diagrams', *Deep-Sea Res.*, **18**, 789–802.

Pickard, G. L., 1956. 'Physical features of British Columbia inlets', *Trans. Roy. Soc. Canada*, **50**, 47–58.

Pickard, G. L., 1961. 'Oceanographic features of inlets in the British Columbia mainland coast', *Jour. Fish. Res. Bd. Canada*, **18**, 907–999.

Pickard, G. L., 1971. 'Some physical oceanographic features of inlets of Chili', *Jour. Fish. Res. Bd. Canada*, **28**, 1077–1106.

Pickard, G. L. and Rodgers, K., 1959. 'Current measurements in Knight Inlet, British Columbia', *Jour. Fish. Res. Bd. Canada*, **16**, 635–678.

Preddy, W. S., 1954. 'The mixing and movement of water in the estuary of the Thames', *Jour. Marine Biol. Assoc. U.K.*, **33**, 645–662.

Price, W. A. and Kendrick, M. P., 1963. 'Field and model investigations into the reason for siltation in the Mersey estuary', *Proc. Inst. Civil Eng.*, **24**, 473–518.

Pritchard, D. W., 1952a. 'Salinity distribution and circulation in the Chesapeake Bay estuaries system', *Jour. Marine Res.*, **11**, 106–123.

Pritchard, D. W., 1952b. 'Estuarine hydrography', *Advan. Geophy.* **1**, 243–280.

Pritchard, D. W., 1954. 'A study of the salt balance in a coastal plain estuary', *Jour. Marine Res.*, **13**, 133–144.

Pritchard, D. W., 1955. 'Estuarine circulation patterns', *Proc. Amer. Soc. Civil Eng.*, **81**, No. 717.

Pritchard, D. W., 1956. 'The dynamic structure of a coastal plain estuary', *Jour. Marine Res.*, **15**, 33–42.

Pritchard, D. W., 1958. 'The equations of mass continuity and salt continuity in estuaries', *Jour. Marine Res.*, **17**, 412–423.

Pritchard, D. W., 1967. 'Observations of circulation in coastal plain estuaries', in *Estuaries* (ed. G. H. Lauff), Amer. Assoc. Adv. Sci.

Pritchard, D. W., 1969. 'Dispersion and flushing of pollutants in estuaries', *Proc. Amer. Soc. Civil Eng.*, **95**, HY1, 115–124.

Pritchard, D. W. and Burt, W. V., 1951. 'An inexpensive and rapid technique for obtaining current profiles in estuarine waters', *Jour. Marine Res.*, **10**, 180–189.

Pritchard, D. W. and Carpenter, J. H., 1960. 'Measurements of turbulent diffusion in estuarine and inshore waters', *Bull. Int. Assoc. Sci. Hydrol.*, **20**, 37–50.

Pritchard, D. W. and Kent, R. E., 1953. 'The reduction and analysis of data from the James River Operation Oyster Spot', Tech Rept. VI, Ref 53–12, Chesapeake Bay Inst., Johns Hopkins Univ.

Pritchard, D. W. and Kent, R. E., 1956. 'A method for determining mean longitudinal velocities in a coastal plain estuary', *Jour. Marine Res.*, **15**, 81–91.

Pyatt, E. E., 1964. 'On determining pollutant distribution in tidal estuaries', *Geol. Surv. Water Supp. Paper* 1586-F.

Rattray, M. Jr., 1967. 'Some aspects of the dynamics of circulations in fjords', in *Estuaries* (ed. G. H. Lauff), Amer. Assoc. Adv. Sci.

Rattray, M. Jr. and Hansen, D. V., 1962. 'A similarity solution for circulation in an estuary', *Jour. Marine Res.*, **20**, 121–133.

Saelen, O. H., 1967. 'Some features of the hydrography of Norwegian fjords', in *Estuaries* (ed. G. H. Lauff), Amer. Assoc. Adv. Sci.

Simmons, H. B., 1955. 'Some effects of upland discharge on estuarine hydraulics', *Proc. Amer. Soc. Civil Eng.*, **81**, No. 792.

Sternberg, R. W., 1968. 'Friction factors in tidal channels with differing bed roughness', *Marine Geol.*, **6**, 243–260.

Stewart, R. W., 1958. 'A note on the dynamic balance in estuarine circulation', *Jour. Marine Res.*, **16**, 34–39.

Stommel, H., 1953a. 'Computation of pollution in a vertically mixed estuary', *Sewage Indust. Wastes*, **25**, 1065–1071.

Stommel, H., 1953b. 'The role of density currents in estuaries', *Proc. Minn. Int. Hydrol. Conr.*

Stommel, H. and Farmer, H. G., 1952. 'Abrupt change in width in two-layer open channel flow', *Jour. Marine Res.*, **11**, 205–214.

Stommel, H. and Farmer, H. G., 1953. 'Control of salinity in an estuary by a transition', *Jour. Marine Res.*, **12**, 13–20.

Taylor, G., 1954. 'The dispersion of matter in turbulent flow through a pipe', *Proc. Roy. Soc.*, **A223**, 446–468.

Tully, J. P., 1949. 'Oceanography and prediction of pulp mill pollution in Alberni Inlet', *Bull. Fish. Res. Bd. Canada*, **83**, 169.

Tully, J. P., 1958. 'On structure, entrainment and transport in estuarine embayments', *Jour. Marine Res.*, **17**, 523–535.

Wright, L. D., 1971. 'Hydrography of South Pass, Mississippi River', *Proc. Amer. Soc. Civil Eng.*, **97**, WW3, 491–504.

Author Index

135

Subject Index